Geometry

LARSON
BOSWELL
STIFF

Applying • Reasoning • Measuring

Chapter 12
Resource Book

The Resource Book contains the wide variety
of blackline masters available for Chapter 12.
The blacklines are organized by lesson. Included
are support materials for the teacher as well as
practice, activities, applications, and assessment
resources.

McDougal Littell
A HOUGHTON MIFFLIN COMPANY
Evanston, Illinois • Boston • Dallas

Contributing Authors

The authors wish to thank the following individuals for their contributions to the Chapter 12 Resource Book.

Eric J. Amendola
Karen Collins
Michael Downey
Patrick M. Kelly
Edward H. Kuhar
Lynn Lafferty
Dr. Frank Marzano
Wayne Nirode
Dr. Charles Redmond
Paul Ruland

ISBN: 0-618-02075-6

9-VEI- 04

Contents

12 *Surface Area and Volume*

Chapter Support	1–8
12.1 Exploring Solids	9–20
12.2 Surface Area of Prisms and Cylinders	21–33
12.3 Surface Area of Pyramids and Cones	34–50
12.4 Volume of Prisms and Cylinders	51–64
12.5 Volume of Pyramids and Cones	65–77
12.6 Surface Area and Volume of Spheres	78–93
12.7 Similar Solids	94–106
Review and Assess	107–120
Resource Book Answers	A1–A10

Contents

CHAPTER SUPPORT MATERIALS

Tips for New Teachers	p. 1	Prerequisite Skills Review	p. 5
Parent Guide for Student Success	p. 3	Strategies for Reading Mathematics	p. 7

LESSON MATERIALS

	12.1	12.2	12.3	12.4	12.5	12.6	12.7
Lesson Plans (Reg. & Block)	p. 9	p. 21	p. 34	p. 51	p. 65	p. 78	p. 94
Warm-Ups & Daily Quiz	p. 11	p. 23	p. 36	p. 53	p. 67	p. 80	p. 96
Activity Support Masters		p. 24					
Alternative Lesson Openers	p. 12	p. 25	p. 37	p. 54	p. 68	p. 81	p. 97
Tech. Activities & Keystrokes			p. 38	p. 55		p. 82	
Practice Level A	p. 13	p. 26	p. 41	p. 57	p. 69	p. 86	p. 98
Practice Level B	p. 14	p. 27	p. 42	p. 58	p. 70	p. 87	p. 99
Practice Level C	p. 15	p. 28	p. 43	p. 59	p. 71	p. 88	p. 100
Reteaching and Practice	p. 16	p. 29	p. 44	p. 60	p. 72	p. 89	p. 101
Catch-Up for Absent Students	p. 18	p. 31	p. 46	p. 62	p. 74	p. 91	p. 103
Coop. Learning Activities							p. 104
Interdisciplinary Applications	p. 19		p. 47		p. 75		p. 105
Real-Life Applications		p. 32		p. 63		p. 92	
Math and History Applications			p. 48				
Challenge: Skills and Appl.	p. 20	p. 33	p. 49	p. 64	p. 76	p. 93	p. 106

REVIEW AND ASSESSMENT MATERIALS

Quizzes	p. 50, p. 77	Alternative Assessment & Math Journal	p. 115
Chapter Review Games and Activities	p. 107	Project with Rubric	p. 117
Chapter Test (3 Levels)	p. 108	Cumulative Review	p. 119
SAT/ACT Chapter Test	p. 114	Resource Book Answers	p. A1

Contents

Descriptions of Resources

This Chapter Resource Book is organized by lessons within the chapter in order to make your planning easier. The following materials are provided:

Tips for New Teachers These teaching notes provide both new and experienced teachers with useful teaching tips for each lesson, including tips about common errors and inclusion.

Parent Guide for Student Success This guide helps parents contribute to student success by providing an overview of the chapter along with questions and activities for parents and students to work on together.

Prerequisite Skills Review Worked-out examples are provided to review the prerequisite skills highlighted on the Study Guide page at the beginning of the chapter. Additional practice is included with each worked-out example.

Strategies for Reading Mathematics The first page teaches reading strategies to be applied to the current chapter and to later chapters. The second page is a visual glossary of key vocabulary.

Lesson Plans and Lesson Plans for Block Scheduling This planning template helps teachers select the materials they will use to teach each lesson from among the variety of materials available for the lesson. The block-scheduling version provides additional information about pacing.

Warm-Up Exercises and Daily Homework Quiz The warm-ups cover prerequisite skills that help prepare students for a given lesson. The quiz assesses students on the content of the previous lesson. (Transparencies also available)

Activity Support Masters These blackline masters make it easier for students to record their work on selected activities in the Student Edition.

Alternative Lesson Openers An engaging alternative for starting each lesson is provided from among these four types: *Application, Activity, Geometry Software,* or *Visual Approach.* (Color transparencies also available)

Technology Activities with Keystrokes Keystrokes for Geometry software and calculators are provided for each Technology Activity in the Student Edition, along with alternative Technology Activities to begin selected lessons.

Practice A, B, and C These exercises offer additional practice for the material in each lesson, including application problems. There are three levels of practice for each lesson: A (basic), B (average), and C (advanced).

Contents

Reteaching with Practice These two pages provide additional instruction, worked-out examples, and practice exercises covering the key concepts and vocabulary in each lesson.

Quick Catch-Up for Absent Students This handy form makes it easy for teachers to let students who have been absent know what to do for homework and which activities or examples were covered in class.

Cooperative Learning Activities These enrichment activities apply the math taught in the lesson in an interesting way that lends itself to group work.

Interdisciplinary Applications/Real-Life Applications Students apply the mathematics covered in each lesson to solve an interesting interdisciplinary or real-life problem.

Math and History Applications This worksheet expands upon the Math and History feature in the Student Edition.

Challenge: Skills and Applications Teachers can use these exercises to enrich or extend each lesson.

Quizzes The quizzes can be used to assess student progress on two or three lessons.

Chapter Review Games and Activities This worksheet offers fun practice at the end of the chapter and provides an alternative way to review the chapter content in preparation for the Chapter Test.

Chapter Tests A, B, and C These are tests that cover the most important skills taught in the chapter. There are three levels of test: A (basic), B (average), and C (advanced).

SAT/ACT Chapter Test This test also covers the most important skills taught in the chapter, but questions are in multiple-choice and quantitative-comparison format. (See *Alternative Assessment* for multi-step problems.)

Alternative Assessment with Rubrics and Math Journal A journal exercise has students write about the mathematics in the chapter. A multi-step problem has students apply a variety of skills from the chapter and explain their reasoning. Solutions and a 4-point rubric are included.

Project with Rubric The project allows students to delve more deeply into a problem that applies the mathematics of the chapter. Teacher's notes and a 4-point rubric are included.

Cumulative Review These practice pages help students maintain skills from the current chapter and preceding chapters.

LESSON 12.1

TEACHING TIP Have models similar to those in the summary table on page 719 available for students to look at and touch. Wood or plastic models may be available in your math, art, or science departments. Your science department may have crystals similar to those in Exercises 39 through 41 on page 725 that you could use for demonstrations.

TEACHING TIP You could also make some demonstration models out of paper, regular drinking straws or flex straws, or toothpicks and miniature marshmallows or gumdrops. Modeling clay could be used, also. You could make or have students make as many models as you like and use them to substantiate Euler's Theorem on page 721. Students' spatial perceptions improve when they can examine real three-dimensional models and learn how to draw the models' two-dimensional images.

LESSON 12.2

TEACHING TIP When describing the lateral faces of a prism as a parallelogram, remind students of the other quadrilaterals that are also parallelograms. It is possible for a prism to have lateral faces that are in the shape of a square, a rectangle, or a rhombus, as well as the standard parallelogram shape.

TEACHING TIP Show students how to draw solid models, if you haven't done so yet, and refer them to the Study Tip on page 728 as a guide. Explain about using dashed lines for the part of the solid that cannot be seen from the front view. Using dashed lines helps to give the viewer a sense of depth for the solid and a sense of the part that is out of view.

INCLUSION Consider having nets of a few solids available to help students with limited English proficiency. Make nets like those on pages 729 and 732. This should help students to make a connection between the theorems and the models for the surface area of a right prism and the surface area of a right cylinder.

COMMON ERROR Encourage students to first write a surface area formula for a solid when problem solving. They should check their formulas for any solid (simple or complex) to make sure all surfaces are included.

LESSON 12.3

COMMON ERROR Students may confuse slant height with height for a pyramid. Stress that the slant height is the altitude of a lateral face. Consider having them draw diagrams that include both the pyramid and the two-dimensional right triangle as in Example 1 on page 735. Drawing and labeling both diagrams should help them avoid confusion.

TEACHING TIP Refer to the Study Tip and the diagrams at the top of page 736 to help students draw a net and analyze the surface area for different types of pyramids. Again, encourage students to write the formulas for surface area before beginning the algebraic solution. They should always check to make sure all surfaces have been included.

LESSON 12.4

TEACHING TIP Consider having some type of model available for demonstration similar to the model in Example 1 on page 743. You could use a set of smaller boxes and a larger one, blocks of wood and their container, or a box of sugar cubes. There are many possible models that can give students a real example of volume measuring a finite space.

TEACHING TIP If your school has a wood shop, you could ask the teacher or a student to make you a set of models like those on page 744 to demonstrate Cavalieri's Principle. You might have to help them with calculations to assure that all three have the same height and cross sections of equal area.

COMMON ERROR Students may leave off the units of measure when finding volume or possibly write them with square units instead of cubic units. Remind students to always check that their final answers include the correct units of measure.

They also need to be reminded that all measurements should be in the same units (all in feet or all in centimeters, for example) when first applying any surface area or volume formulas.

LESSON 12.5

TEACHING TIP Let students know whether you want them to leave answers in terms of π or to give approximate decimal answers for problems like those illustrated in Examples 2 and 3 on page 753.

COMMON ERROR Students have a tendency to think that the volume of a pyramid and a cone are $\frac{1}{2}$ the volume of a prism, when all have the same height. It is a matter of their visual perception when they look at models or diagrams like those on page 752. Assure them that the relationship is $\frac{1}{3}$. This can be demonstrated by using hollow models and filling them with sand (sugar, salt, or water may work as well). Practice before you attempt the demonstration in class.

LESSON 12.6

TEACHING TIP Show students how to draw a sphere by first drawing a circle and then adding an ellipse. Similar to the diagrams on page 759, half of the ellipse should be solid and half should be dashed to create a three-dimensional perspective.

TEACHING TIP Finding the volume of a sphere given its radius will involve cubing the radius. Review cubing with students. In Example 4 on page 761, the problem involves finding a cube root. Students may need help using their calculators to find a cube root. If their calculators do not have a cube root function, consider letting them leave their answer in cube root radical form.

LESSON 12.7

TEACHING TIP Have students recall properties of similar polygons and then outline properties of similar solids, as indicated at the top of page 766. Understanding the similarities between two-dimensional and three-dimensional figures should be an asset to students. Recall that scale factors or ratios should always be given in simplest form.

TEACHING TIP Theorem 12.13 on page 767 indicates that corresponding areas have a ratio of $a^2 : b^2$. Emphasize that this ratio of areas is accurate for total surface areas, as well as for corresponding lateral areas or base areas.

Outside Resources

BOOKS/PERIODICALS

Davis, Elwyn H. "Area of Spherical Triangles." *Mathematics Teacher* (February 1999); pp. 150–153.

Naylor, Michael. "The Amazing Octacube." *Mathematics Teacher* (February, 1999); pp. 102–104.

ACTIVITIES/MANIPULATIVES

Lenart, Istvan. *Lenart Sphere Basic Set.* Construction materials; includes a 15 piece set plus instruction booklet. Berkeley, CA; Key Curriculum Press.

VIDEOS

The Platonic Solids Video. Computer animation illustrates properties. Activity book with activities and explorations also available. Berkeley, CA; Key Curriculum Press.

NAME _____ DATE _____

Parent Guide for Student Success

For use with Chapter 12

Chapter Overview One way that you can help your student succeed in Chapter 12 is by discussing the lesson goals in the chart below. When a lesson is completed, ask your student to interpret the lesson goals for you and to explain how the mathematics of the lesson relates to one of the key applications listed in the chart.

Lesson Title	Lesson Goals	Key Applications
12.1: Exploring Solids	Use properties of polyhedra. Use Euler's Theorem in real-life situations.	• Soccer Ball • Cooking • Crystals
12.2: Surface Area of Prisms and Cylinders	Find the surface area of a prism and of a cylinder.	• Packaging • Wax Cylinder Records • Cake Design
12.3: Surface Area of Pyramids and Cones	Find the surface area of a pyramid and of a cone.	• Memphis Pyramid Arena • Lampshades • Squirrel Barrier
12.4: Volume of Prisms and Cylinders	Use volume postulates. Find the volume of prisms and of cylinders in real life.	• Ennis-Brown House • Candles • Aquarium Tank
12.5: Volume of Pyramids and Cones	Find the volume of a pyramid and cones. Find the volume of pyramids and of cones in real life.	• Nautical Prisms • Automatic Feeder • Volcanoes
12.6: Surface Area and Volume of Spheres	Find the surface area of a sphere. Find the volume of a sphere in real life.	• Ball Bearings • Planets • Spheres in Architecture
12.7: Similar Solids	Find and use the scale factor of similar solids. Use similar solids to solve real-life problems.	• Meteorology • Swimming Pools • Civil Rights Institute

Study Strategy

Generalizing Formulas is the study strategy featured in Chapter 12 (see page 718). Topics like surface area and volume involve a lot of formulas. Rather than remembering every formula individually, encourage your student to find underlying concepts that link many formulas together. Then, your student will only have to remember the concept instead of all the formulas.

NAME _____ DATE _____

Parent Guide for Student Success

For use with Chapter 12

Key Ideas Your student can demonstrate understanding of key concepts by working through the following exercises with you.

Lesson	Exercise
12.1	A pyramid has 7 vertices and 12 edges. How many faces does it have? What is the shape of the pyramid's base?
12.2	A cereal box measures 8 centimeters by 20 centimeters by 28 centimeters. Not counting overlap, how much cardboard would it take to make the box?
12.3	Find the surface area of a cone with radius 9 inches and height 12 inches.
12.4	A cylindrical silo has a radius of 3 meters and a height of 10 meters. How much grain can it hold?
12.5	Find the volume of a pyramid with a square base that is 70 feet on a side and has a height of 90 feet.
12.6	Earth's radius is about 4000 miles. What is the surface area of Earth? The Pacific Ocean covers approximately 64,186,300 square miles. What percent is this of Earth's surface?
12.7	A cone has a volume of 81π cubic meters. A similar cone has a volume of 192π cubic meters. Find their scale factor.

Home Involvement Activity

You will need: Grid paper, scissors, tape

Directions: Cut a 12 by 12 square out of the grid paper. Cut a 1 by 1 square out of each corner and fold to form an open-top box. Find the volume of the box. Repeat by cutting a 2 by 2 square out of each corner, then a 3 by 3 square and so on. What are the integer dimensions of the box that maximizes volume while using the 12 by 12 square? Start with a different rectangular shape and try to guess what dimensions will maximize volume. Calculate to check.

Answers

12.1: 7 faces; hexagon **12.2:** 1,888 cm² **12.3:** about 678.6 in.² **12.4:** 282.7 m³ **12.5:** 147,000 ft³ **12.6:** 201,061,929 mi²; about 32% **12.7:** 3 : 4 **Home Involvement Activity:** An 8 by 8 by 2 box has the maximum volume.

NAME _____ DATE _____

Prerequisite Skills Review

For use before Chapter 12

EXAMPLE 1 *Finding Scale Factors of Similar Polygons*

Find the scale factor of the similar polygons.

a.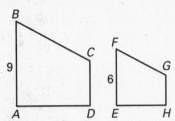

b.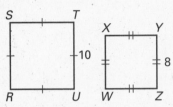

SOLUTION

a. Because $ABCD \sim EFGH$, the corresponding side lengths are proportional.

$$\frac{AB}{EF} = \frac{9}{6} = \frac{3}{2}$$

The scale factor is 3:2.

b. Because $RSTU \sim WXYZ$, the corresponding side lengths are proportional.

$$\frac{TU}{YZ} = \frac{10}{8} = \frac{5}{4}$$

The scale factor is 5:4.

Exercises for Example 1

Find the scale factor of the similar polygons.

1.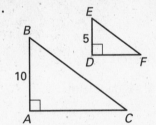

2.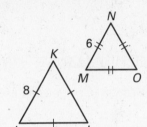

3.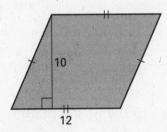

EXAMPLE 2 *Finding the Areas of Quadrilaterals*

Find the area of the quadrilateral.

a.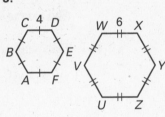

b.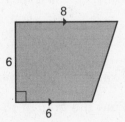

SOLUTION

a. $A = bh$ Area of parallelogram

$A = 10(12)$ Substitute.

$A = 120$ units^2 Simplify.

NAME _____ DATE _____

Prerequisite Skills Review

For use before Chapter 12

b. $A = \dfrac{h(b_1 + b_2)}{2}$ Area of trapezoid

  $A = \dfrac{6(6 + 8)}{2}$ Substitute.

  $A = \dfrac{6(14)}{2}$ Simplify.

  $A = 42$ units^2 Simplify.

Exercises for Example 2

Find the area of the quadrilateral.

4.
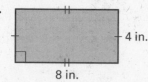
4 in.

8 in.

5.
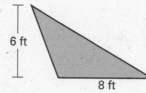
6 ft

8 ft

6.
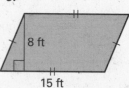
8 ft

15 ft

EXAMPLE 3 *Find the Area of Regular Polygons*

Find the area of the regular polygon
to the nearest tenth.

SOLUTION

The measure of central $\angle ABC$ is $\dfrac{1}{6} \cdot 360°$, or $60°$.

The apothem bisects $\angle B$, so you can use the trigonometric

ratio $\tan 30° = \dfrac{\dfrac{s}{2}}{2\sqrt{3}} = \dfrac{s}{4\sqrt{3}}$. Solving for s yields

$s = 4\sqrt{3} \tan 30° = 4\sqrt{3} \cdot \dfrac{\sqrt{3}}{3} = 4.$

Use the formula for the area of a regular polygon.

$A = \dfrac{1}{2}a \cdot ns = \dfrac{1}{2}(2\sqrt{3})(6s) = \dfrac{1}{2}(2\sqrt{3})(6 \cdot 4) = 24\sqrt{3} \approx 41.6 \text{ cm}^2$

$2\sqrt{3}$ cm

Exercises for Example 3

Find the area of the regular polygon to the nearest tenth.

7.
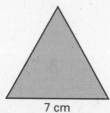
7 cm

8.
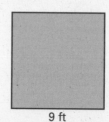
9 ft

9.
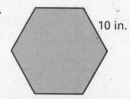
10 in.

Geometry
Chapter 12 Resource Book

NAME _____ DATE _____

Strategies for Reading Mathematics

For use with Chapter 12

Strategy: Interpreting Diagrams of Three-Dimensional Figures

In order to show three-dimensional figures on a flat piece of paper, they must be drawn as two-dimensional diagrams. Solid and dashed edges are used to make the figures shown in the diagrams appear to be three-dimensional.

Dashed lines are used to show edges at the back or on the bottom of the three-dimensional figure. Any face with at least one dashed edge would not be visible from the front.

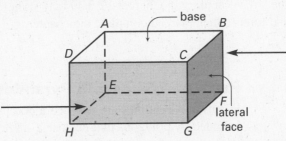

Solid edges are used to show faces at the front or on the top of the three-dimensional figure. A face with all solid edges would be visible from the front.

STUDY TIP

Understanding Diagrams of Three-Dimensional Figures

When you view a two-dimensional diagram of a three-dimensional figure, think about what you would see if you were looking at the actual figure. The solid edges are the ones you would actually see. The dashed edges are the ones that would be hidden.

Questions

1. Refer to the prism above. Name the base that is on the top and the lateral faces that would be visible to a viewer looking at the front of the prism.

2. Refer to the prism above. Name the base that is on the bottom and the lateral faces that would be hidden from a viewer looking at the front of the prism.

3. Suppose you are looking at a three-dimensional model of the prism shown above. What edges would you see? How do you know? What edges would be hidden? How do you know?

4. A two-dimensional diagram of a pyramid is shown at the right. Name the faces that are at the front and the faces that are at the back and on the bottom of the three-dimensional figure.

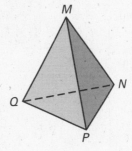

Strategies for Reading Mathematics

For use with Chapter 12

Visual Glossary

The Study Guide on page 718 lists the key vocabulary for Chapter 12 as well as review vocabulary from previous chapters. Use the page references on page 718 or the Glossary in the textbook to review key terms from prior chapters. Use the visual glossary below to help you understand some of the key vocabulary in Chapter 12. You may want to copy these diagrams into your notebook and refer to them as you complete the chapter.

GLOSSARY

prism (p. 728) A polyhedron with two congruent *bases* that lie in parallel planes. The *lateral faces* are parallelograms that connect the bases. The height is the perpendicular distance between its bases.

pyramid (p. 735) A polyhedron in which the base is a polygon and the *lateral faces* are triangles with a common *vertex*. The *height* is the perpendicular distance between the base and the vertex.

cylinder (p. 730) A solid with congruent circular bases that lie in parallel planes. The *height* of a cylinder is the perpendicular distance between its bases. The radius of the base is also the *radius* of the cylinder.

circular cone (p. 737) A solid with a circular *base* and a *vertex* that is not in the same plane as the base. The *lateral surface* consists of all segments that connect the vertex with the edge of the base. The *height* is the perpendicular distance between the vertex and the plane that contains the base.

Parts of Prisms and Pyramids

Prisms and pyramids are polyhedra because they are solids that are bounded by polygons that enclose a single region of space. Their important parts are shown in the diagrams below.

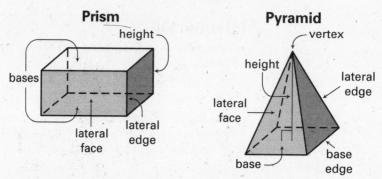

Parts of Cylinders and Cones

Cylinders and cones are solids with curved surfaces. Their important parts are shown in the diagrams below.

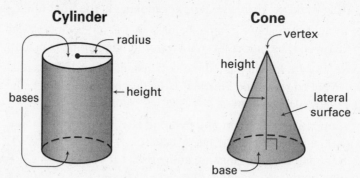

LESSON 12.1

TEACHER'S NAME _____ CLASS _____ ROOM _____ DATE _____

Lesson Plan

1-day lesson (See *Pacing the Chapter*, TE pages 716C–716D)   **For use with pages 719–726**

GOALS   1. **Use properties of polyhedra.**
2. **Use Euler's Theorem in real-life situations.**

State/Local Objectives _____

_____

✓ **Check the items you wish to use for this lesson.**

STARTING OPTIONS
____ Prerequisite Skills Review: CRB pages 5–6
____ Strategies for Reading Mathematics: CRB pages 7–8
____ Homework Check: TE page 702: Answer Transparencies
____ Warm-Up or Daily Homework Quiz: TE pages 719 and 704, CRB page 11, or Transparencies

TEACHING OPTIONS
____ Motivating the Lesson: TE page 720
____ Lesson Opener (Activity): CRB page 12 or Transparencies
____ Examples 1–6: SE pages 719–722
____ Extra Examples: TE pages 720–722 or Transparencies
____ Closure Question: TE page 722
____ Guided Practice Exercises: SE page 723

APPLY/HOMEWORK
Homework Assignment
____ Basic   10–30 even, 32–35, 43–52, 54, 55, 60–70 even
____ Average   10–30 even, 32–35, 42–52, 54, 55, 60–70 even
____ Advanced   10–30 even, 32–35, 42–52, 54–59, 60–70 even

Reteaching the Lesson
____ Practice Masters: CRB pages 13–15 (Level A, Level B, Level C)
____ Reteaching with Practice: CRB pages 16–17 or Practice Workbook with Examples
____ Personal Student Tutor

Extending the Lesson
____ Applications (Interdisciplinary): CRB page 19
____ Challenge: SE page 726; CRB page 20 or Internet

ASSESSMENT OPTIONS
____ Checkpoint Exercises: TE pages 720–722 or Transparencies
____ Daily Homework Quiz (12.1): TE page 726, CRB page 23, or Transparencies
____ Standardized Test Practice: SE page 726; TE page 726; STP Workbook; Transparencies

Notes _____

_____

_____

Lesson 12.1

TEACHER'S NAME _____ CLASS _____ ROOM _____ DATE _____

Lesson Plan for Block Scheduling

Half-day lesson (See *Pacing the Chapter,* TE pages 716C–716D)   For use with pages 719–726

GOALS 1. **Use properties of polyhedra.**
2. **Use Euler's Theorem in real-life situations.**

State/Local Objectives _____

_____

_____

✓ **Check the items you wish to use for this lesson.**

STARTING OPTIONS

_____ Prerequisite Skills Review: CRB pages 5–6
_____ Strategies for Reading Mathematics: CRB pages 7–8
_____ Homework Check: TE page 702: Answer Transparencies
_____ Warm-Up or Daily Homework Quiz: TE pages 719 and
        704, CRB page 11, or Transparencies

TEACHING OPTIONS

_____ Motivating the Lesson: TE page 720
_____ Lesson Opener (Activity): CRB page 12 or Transparencies
_____ Examples 1–6: SE pages 719–722
_____ Extra Examples: TE pages 720–722 or Transparencies
_____ Closure Question: TE page 722
_____ Guided Practice Exercises: SE page 723

APPLY/HOMEWORK

Homework Assignment

_____ Block Schedule: 10–30 even, 32–35, 42–52, 54, 55, 60–70 even

Reteaching the Lesson

_____ Practice Masters: CRB pages 13–15 (Level A, Level B, Level C)
_____ Reteaching with Practice: CRB pages 16–17 or Practice Workbook with Examples
_____ Personal Student Tutor

Extending the Lesson

_____ Applications (Interdisciplinary): CRB page 19
_____ Challenge: SE page 726; CRB page 20 or Internet

ASSESSMENT OPTIONS

_____ Checkpoint Exercises: TE pages 720–722 or Transparencies
_____ Daily Homework Quiz (12.1): TE page 726, CRB page 23, or Transparencies
_____ Standardized Test Practice: SE page 726; TE page 726; STP Workbook; Transparencies

Notes _____

_____

_____

CHAPTER PACING GUIDE	
Day	**Lesson**
1	Assess Ch. 11; **12.1 (all)**
2	12.2 (all); 12.3 (begin)
3	12.3 (end); 12.4 (begin)
4	12.4 (end); 12.5 (begin)
5	12.5 (end); 12.6 (all)
6	12.7 (all)
7	Review Ch. 12; Assess Ch. 12

NAME _____ DATE _____

WARM-UP EXERCISES

For use before Lesson 12.1, pages 719–726

State the number of sides of each polygon.

1. nonagon

2. hexagon

3. decagon

4. dodecagon

5. n-gon

DAILY HOMEWORK QUIZ

For use after Lesson 11.6, pages 699–706

1. Find the probability that a point chosen on $\overline{AD}$ is on $\overline{BC}$.

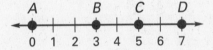

2. Find the probability that a randomly chosen point in the figure lies in the shaded region.

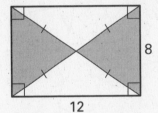

3. Find the value of x so that the probability of the spinner landing on a grey sector is $\dfrac{1}{9}$.

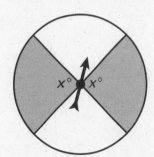

NAME _____ DATE _____

Activity Lesson Opener

For use with pages 719–726

SET UP: Work with a partner or in a group.

YOU WILL NEED: • construction paper • scissors • tape

Cut out and trace around the polygons at the bottom of the page to draw an enlargement of each figure on construction paper. Cut along solid lines, fold along dashed lines, and use tape to create a three-dimensional object. Tell how many and what type of polygons form each object.

1.
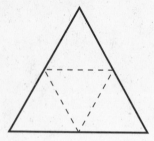

2.
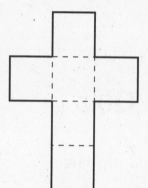

3.
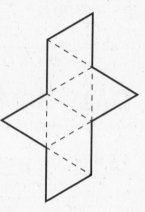

4.
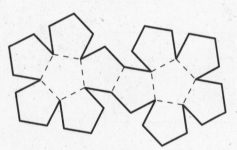

5.
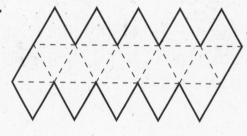

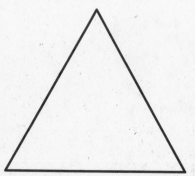

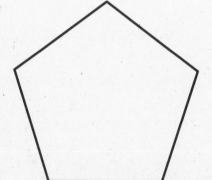

Geometry
Chapter 12 Resource Book

NAME _____ DATE _____

Practice A

For use with pages 719–726

Tell whether the solid is a polyhedron. Explain your reasoning.

1.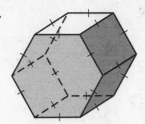

2.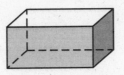

3.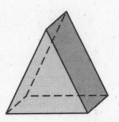

Count the number of faces, vertices, and edges of the polyhedron.

4.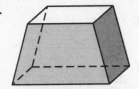

5.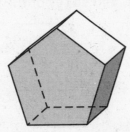

6.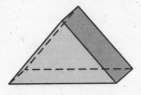

Decide whether the polyhedron is regular and/or convex. Explain.

7.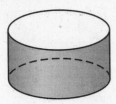

8.

9.

In Exercises 10–13, use the figure shown which represents a barn.

10. How many faces does the barn have?

11. How many vertices does the barn have?

12. How many edges does the barn have?

13. Do your results satisfy Euler's Theorem?

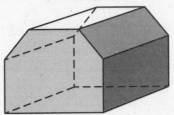

In Exercises 14–19, use the figure shown which represents a piece of cake.

14. How many faces does the piece of cake have?

15. How many vertices does the piece of cake have?

16. How many edges does the piece of cake have?

17. Do your results satisfy Euler's Theorem?

18. Make a sketch to show how a cross section of the piece of cake could be a triangle.

19. Make a sketch to show how a cross section of the piece of cake could be a rectangle.

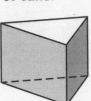

NAME _____ DATE _____

Practice B

For use with pages 719–726

Tell whether the solid is a polyhedron. Explain your reasoning.

1.

2.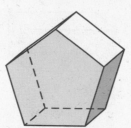

3.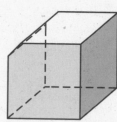

Count the number of faces, vertices, and edges of the polyhedron.

4.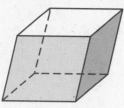

5.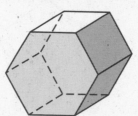

6.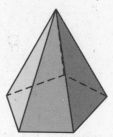

Decide whether the polyhedron is regular and/or convex. Explain.

7.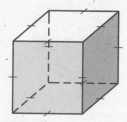

8.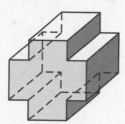

9.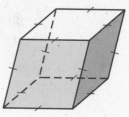

Use Euler's Theorem to find the unknown number.

10. Faces: _____?_____

    Vertices: 6

    Edges: 12

11. Faces: 5

    Vertices: _____?_____

    Edges: 8

12. Faces: 7

    Vertices: 10

    Edges: _____?_____

Describe the cross section.

13.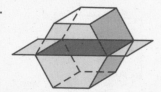

14.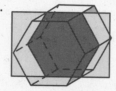

15.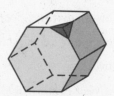

16. Draw a cube. Sketch an example of how the cross section could be

    a. a square.   **b.** a rectangle.   **c.** a triangle.   **d.** a trapezoid.

Practice C

For use with pages 719–726

Tell whether the solid is a polyhedron. Explain your reasoning.

1.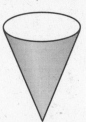

2.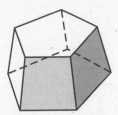

3.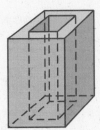

Count the number of faces, vertices, and edges of the polyhedron. Verify that the results satisfy Euler's Theorem.

4.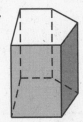

5.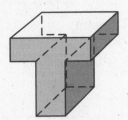

6.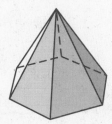

Determine whether the statement is *true* or *false*.

7. A polyhedron can have a circular face.　　**8.** Every regular polyhedron is convex.

9. The cross section of a tetrahedron could be a square.

10. The cross section of a cube could be an equilateral triangle.

11. A polyhedron always has more edges than faces and vertices combined.

12. A polyhedron can have exactly 4 faces and exactly 4 edges.

Describe the cross section shown.

13.

14.

15.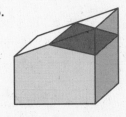

Calculate the number of vertices of the solid using the given information.

16. 12 faces;
all pentagons

17. 14 faces; 8 triangles
and 6 octagons

18. 26 faces; 18 squares
and 8 triangles

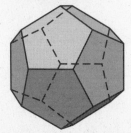

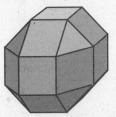

NAME _____ DATE _____

Reteaching with Practice

For use with pages 719–726

GOAL Use properties of polyhedra and use Euler's Theorem

VOCABULARY

A **polyhedron** is a solid that is bounded by polygons that enclose a single region of space.

The polygons that a polyhedron is bounded by are called **faces**.

An **edge** of a polyhedron is a line segment formed by the intersection of two faces.

A **vertex** of a polyhedron is a point where three or more edges meet.

A polyhedron is **regular** if all of its faces are congruent regular polygons.

Theorem 12.1 Euler's Theorem
The number of faces (F), vertices (V), and edges (E) of a polyhedron are related by the formula $F + V = E + 2$.

EXAMPLE 1 *Identifying Polyhedra*

Determine whether each solid is a polyhedron. Explain your reasoning.

a.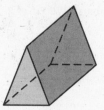

b.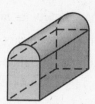

SOLUTION

a. This is a polyhedron. All of its faces are polygons (2 triangles and 3 rectangles), which form a solid enclosing a single region of space.

b. This is not a polyhedron. Some of its faces are not polygons.

Exercises for Example 1

···

Determine whether each solid is a polyhedron. Explain your reasoning.

1.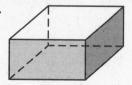

2.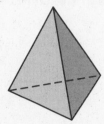

3.

NAME _____ DATE _____

Reteaching with Practice

For use with pages 719–726

EXAMPLE 2 *Analyzing Solids*

For each polyhedron, count the number of faces, vertices, and edges.

a.

b.

SOLUTION

a. The polyhedron has 7 faces, 7 vertices, and 12 edges.

b. The polyhedron has 9 faces, 9 vertices, and 16 edges.

Exercises for Example 2

Count the number of faces, vertices, and edges.

4.

5.

6.
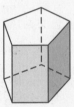

EXAMPLE 3 *Using Euler's Theorem*

Calculate the number of vertices of the solid, given that it has 10 faces, all triangles.

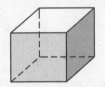

SOLUTION

The 10 triangles alone would have 10(3) = 30 edges. Because each side in the solid is shared by two of these triangles, the total number of edges in the solid is half of this, or 15. Now use Euler's Theorem to find the number of vertices.

$F + V = E + 2$ Write Euler's Theorem.

$10 + V = 15 + 2$ Substitute.

$V = 7$ Solve for V.

Exercise for Example 3

7. Calculate the number of vertices of the solid, given that it has 7 faces; 2 pentagons and 5 rectangles.

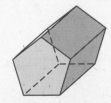

Quick Catch-Up for Absent Students

For use with pages 719–726

The items checked below were covered in class on (date missed) _____

Lesson 12.1: Exploring Solids

_____ **Goal 1:** Use properties of polyhedra. (pp. 719–720)

Material Covered:

_____ Example 1: Identifying Polyhedra

_____ Example 2: Classifying Polyhedra

_____ Student Help: Study Tip

_____ Example 3: Describing Cross Sections

Vocabulary:

polyhedron, p. 719   faces, p. 719

edge, p. 719   vertex, p. 719

regular, p. 720   convex, p. 720

cross section, p. 720

_____ **Goal 2:** Use Euler's Theorem in real-life situations. (pp. 721–722)

Material Covered:

_____ Student Help: Study Tip

_____ Example 4: Using Euler's Theorem

_____ Example 5: Finding the Number of Edges

_____ Example 6: Finding the Number of Vertices

Vocabulary:

Platonic solids, p. 721   tetrahedron, p. 721

octahedron, p. 721   dodecahedron, p. 721

icosahedron, p. 721

_____ Other (specify) _____

_____

Homework and Additional Learning Support

_____ Textbook (specify) <u>pp. 723–726</u>

_____

_____ *Reteaching with Practice* worksheet (specify exercises)_____

_____ *Personal Student Tutor* for Lesson 12.1

NAME _____ DATE _____

Interdisciplinary Application

For use with pages 719–726

Gemstones

SCIENCE   Gemology is the study of semi-precious and precious gem-stones. Gemstones that occur naturally can be classified and identified by their crystal structure. The garnet is a gemstone that is most often red in color, although it does come in a range of colors including bright green. The most common crystal shape for a garnet is a rhombic dodecahedron, a twelve-sided crystal with rhombus faces. Most diamonds and some garnets occur as hexoctahedrons. Lapidaries are gemologists specializing in cutting and polishing stones to maximize brilliance and clarity. Lapidaries take stones in their crystal form and use a variety of techniques to shape the gemstone. Polishing and cutting by a skilled craftsman enhances the beauty of a stone and makes it desirable as a collectible object or piece of jewelry.

In Exercises 1 and 2, use the diagram of the rhombic dodecahedron.

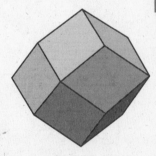

1. Is the common crystal form of a garnet a regular polyhedron? Explain your answer.

2. Given that the rhombic dodecahedron has 12 faces and 14 vertices, how many edges does it have?

In Exercises 3–5, use the diagram of the hexoctahedron.

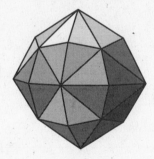

3. The hexoctahedron crystal has 48 triangular faces and 72 edges. Is it a regular polyhedron?

4. How many vertices does a hexoctahedron have?

5. If the hexoctahedron was sliced in half from the top vertex to the bottom vertex, the resulting cross section would be what shape?

Challenge: Skills and Applications

For use with pages 719–726

Euler's Theorem can be extended to polyhedrons with "holes." For a polyhedron with F faces, V vertices, E edges, and n "holes," the equation is $F + V - E = 2 - 2n$. The number $2 - 2n$ is the *Euler characteristic* of the polyhedron. In Exercise 4 below, $n = 3$, so the Euler characteristic is -4.

In Exercises 1–9, count the number of faces, vertices, and edges of the polyhedron. (Assume that the back of the polyhedron matches the front.) Find the Euler characteristic. Check your answers by verifying that
$F + V - E = 2 - 2n.$

1.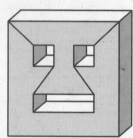

2.

3.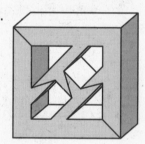

4.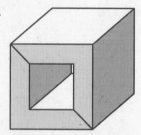

5.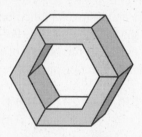

6.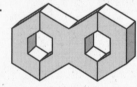

7. A polyhedron has 12 faces, 29 edges, and 1 "hole."
How many vertices does it have?

8. A polyhedron has 23 faces, 49 edges, and 4 "holes."
How many vertices does it have?

9. A polyhedron has 18 vertices, 36 edges, and 2 "holes."
How many faces does it have?

10. A polyhedron has 21 faces, 35 vertices, and 3 "holes."
How many edges does it have?

11. A polyhedron has 26 faces, 43 vertices, and 75 edges.
How many "holes" does it have?

LESSON

12.2

TEACHER'S NAME _____ CLASS _____ ROOM _____ DATE _____

Lesson Plan

1-day lesson (See *Pacing the Chapter,* TE pages 716C–716D)   **For use with pages 727–734**

GOALS   1. **Find the surface area of a prism.**
2. **Find the surface area of a cylinder.**

State/Local Objectives _____

_____

✓ **Check the items you wish to use for this lesson.**

STARTING OPTIONS
____ Homework Check: TE page 723: Answer Transparencies
____ Warm-Up or Daily Homework Quiz: TE pages 728 and 726, CRB page 23, or Transparencies

TEACHING OPTIONS
____ Motivating the Lesson: TE page 729
____ Concept Activity: SE page 727; CRB page 24 (Activity Support Master)
____ Lesson Opener (Visual Approach): CRB page 25 or Transparencies
____ Examples 1–4: SE pages 728–731
____ Extra Examples: TE pages 729–731 or Transparencies; Internet
____ Closure Question: TE page 731
____ Guided Practice Exercises: SE page 731

APPLY/HOMEWORK
Homework Assignment
____ Basic   18–42 even, 45–47, 50–60
____ Average   18–42 even, 44–47, 50–60
____ Advanced   18–42 even, 44–60

Reteaching the Lesson
____ Practice Masters: CRB pages 26–28 (Level A, Level B, Level C)
____ Reteaching with Practice: CRB pages 29–30 or Practice Workbook with Examples
____ Personal Student Tutor

Extending the Lesson
____ Applications (Real-Life): CRB page 32
____ Challenge: SE page 734; CRB page 33 or Internet

ASSESSMENT OPTIONS
____ Checkpoint Exercises: TE pages 729–731 or Transparencies
____ Daily Homework Quiz (12.2): TE page 734, CRB page 36, or Transparencies
____ Standardized Test Practice: SE page 734; TE page 734; STP Workbook; Transparencies

Notes _____

_____

_____

Lesson 12.2

TEACHER'S NAME _____ CLASS _____ ROOM _____ DATE _____

Lesson Plan for Block Scheduling

Half-day lesson (See *Pacing the Chapter*, TE pages 716C–716D)   For use with pages 727–734

GOALS 1. **Find the surface area of a prism.**
2. **Find the surface area of a cylinder.**

State/Local Objectives _____

_____

_____

✓ **Check the items you wish to use for this lesson.**

STARTING OPTIONS

____ Homework Check: TE page 723: Answer Transparencies
____ Warm-Up or Daily Homework Quiz: TE pages 728 and
       726, CRB page 23, or Transparencies

TEACHING OPTIONS

____ Motivating the Lesson: TE page 729
____ Concept Activity: SE page 727; CRB page 24 (Activity Support Master)
____ Lesson Opener (Visual Approach): CRB page 25 or Transparencies
____ Examples 1–4: SE pages 728–731
____ Extra Examples: TE pages 729–731 or Transparencies; Internet
____ Closure Question: TE page 731
____ Guided Practice Exercises: SE page 731

APPLY/HOMEWORK

Homework Assignment (See also the assignment for Lesson 12.3.)

____ Block Schedule: 18–42 even, 44–47, 50–60

Reteaching the Lesson

____ Practice Masters: CRB pages 26–28 (Level A, Level B, Level C)
____ Reteaching with Practice: CRB pages 29–30 or Practice Workbook with Examples
____ Personal Student Tutor

Extending the Lesson

____ Applications (Real-Life): CRB page 32
____ Challenge: SE page 734; CRB page 33 or Internet

ASSESSMENT OPTIONS

____ Checkpoint Exercises: TE pages 729–731 or Transparencies
____ Daily Homework Quiz (12.2): TE page 734, CRB page 36, or Transparencies
____ Standardized Test Practice: SE page 734; TE page 734; STP Workbook; Transparencies

Notes _____

_____

_____

CHAPTER PACING GUIDE	
Day	**Lesson**
1	Assess Ch. 11; 12.1 (all)
2	**12.2 (all)**; 12.3 (begin)
3	12.3 (end); 12.4 (begin)
4	12.4 (end); 12.5 (begin)
5	12.5 (end); 12.6 (all)
6	12.7 (all)
7	Review Ch. 12; Assess Ch. 12

WARM-UP EXERCISES

For use before Lesson 12.2, pages 727–734

Find the area of each polygon.

1. triangle: base = 12 ft, height = 9 ft

2. parallelogram: base = 10 cm, height = 15 cm

3. square: side = 16 in.

4. rectangle: length = 10.2 m, width = 5.5 m

DAILY HOMEWORK QUIZ

For use after Lesson 12.1, pages 719–726

1. Determine whether the statement *A cylinder is a convex polyhedron* is true or false. Explain your reasoning.

2. Describe the cross section.

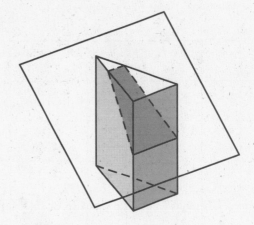

3. Find the number of faces, edges, and vertices of the polyhedron and use them to verify Euler's Theorem.

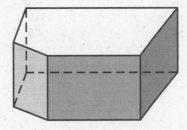

4. A solid has 14 faces: 6 octagons and 8 triangles. How many vertices does it have?

Activity Support Master

For use with page 727

Lesson 12.2

B   F

C

A   D

E

|—— h ——|

NAME ——————————————————— DATE ————

Visual Approach Lesson Opener

For use with pages 728–734

Sketch each solid shown. First draw the two congruent shaded figures. Then connect corresponding vertices, using dashed lines in the back. Identify the two shapes that are faces of the solid, and tell how many of each shape there are.

1.

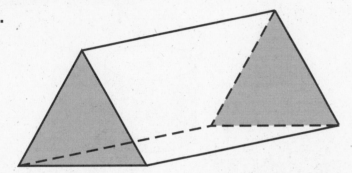

2.

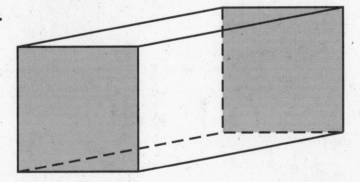

3.

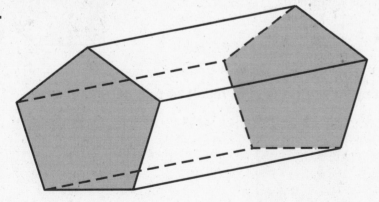

4. Sketch the next solid in the pattern. Identify the two shapes that are faces of the solid, and tell how many of each shape there are.

NAME _____ DATE _____

Practice A
For use with pages 728–734

Use the diagram at the right.

1. Give the mathematical name of the solid.

2. What kind of figure is each base?

3. What kind of figure is each lateral face?

4. How many lateral faces does the solid have?

5. Name three lateral edges.

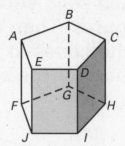

Use the diagram at the right.

6. Give the mathematical name of the solid.

7. What kind of figure is each base?

8. Name a radius of the solid.

9. Name a height of the solid.

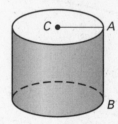

Name the solid that can be folded from the net.

10.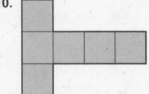

11.

12.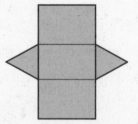

Find the surface area of the right prism.

13.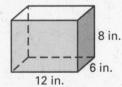

14.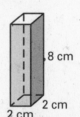

15.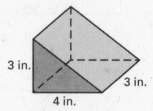

Find the surface area of the right cylinder. Round your result to two decimal places.

16.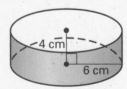

17.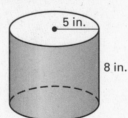

18.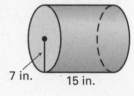

NAME _____   DATE _____

Practice B
For use with pages 728–734

Use the diagram at the right.

1. Give the mathematical name of the solid.

2. What kind of figure is each base?

3. What kind of figure is each lateral face?

4. How many lateral faces does the solid have?

5. Name three lateral edges.

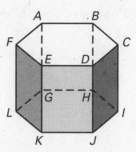

Name the solid that can be folded from the net.

6.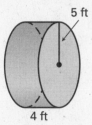

7.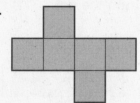

8.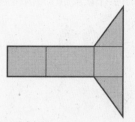

Find the surface area of the right prism. Round your result to two decimal places.

9.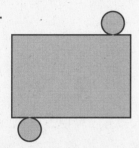

10.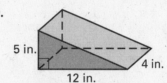

11.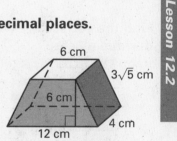

Find the surface area of the right cylinder. Round your result to two decimal places.

12.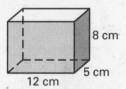

13.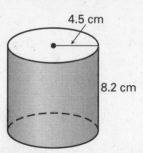

14.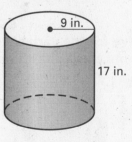

Solve for the variable given the surface area S of the right prism or right cylinder. Round the result to one decimal place.

15. $S = 208 \text{ m}^2$

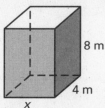

16. $S = 97.5 \text{ cm}^2$

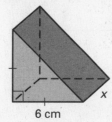

17. $S = 452.4 \text{ in.}^2$

NAME _____ DATE _____

Practice C

For use with pages 728–734

Give the mathematical name of the solid.

1.

2.

3.

4.

Name the solid that can be folded from the net.

5.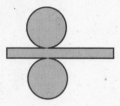

6.

7.

Find the surface area of the right prism or right cylinder. Round your result to two decimal places.

8.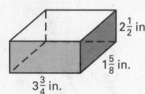
$2\frac{1}{2}$ in.
$1\frac{5}{8}$ in.
$3\frac{3}{4}$ in.

9.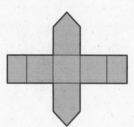
6.4 cm
8.7 cm

10.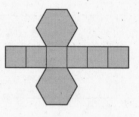
19.6 cm
38.2 cm

Solve for the variable given the surface area *S* of the right prism or right cylinder. Round the result to one decimal place.

11. $S = 363 \text{ m}^2$

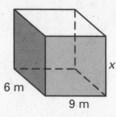
x
6 m
9 m

12. $S = 229.2 \text{ cm}^2$

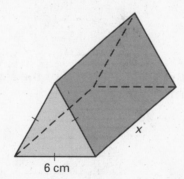
6 cm
x

13. $S = 1206.4 \text{ in.}^2$

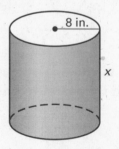
8 in.
x

Lesson 12.2

NAME _____ DATE _____

Reteaching with Practice

For use with pages 728–734

GOAL **Find the surface area of a prism and find the surface area of a cylinder**

VOCABULARY

A **prism** is a polyhedron with two congruent faces, called **bases**, that lie in parallel planes. The other faces are parallelograms formed by connecting the corresponding vertices of the bases and are called **lateral faces**.

In a **right prism**, each lateral edge is perpendicular to both bases.

Prisms that have lateral edges that are not perpendicular to the bases are **oblique prisms**.

The **surface area of a polyhedron** is the sum of the areas of its faces.

The **lateral area of a polyhedron** is the sum of the areas of its lateral faces.

The two-dimensional representation of all of a prism's faces is called a **net**.

A **cylinder** is a solid with congruent circular bases that lie in parallel planes.

A cylinder is called a **right cylinder** if the segment joining the centers of the bases is perpendicular to the bases.

The **lateral area of a cylinder** is the area of its curved surface and is equal to the product of the circumference and the height.

The entire **surface area of a cylinder** is equal to the sum of the lateral area and the areas of the two bases.

Theorem 12.2 Surface Area of a Right Prism
The surface area S of a right prism can be found using the formula $S = 2B + Ph$, where B is the area of a base, P is the perimeter of a base, and h is the height.

Theorem 12.3 Surface Area of a Right Cylinder
The surface area S of a right cylinder is

$$S = 2B + Ch = 2\pi r^2 + 2\pi rh,$$

where B is the area of a base, C is the circumference of a base, r is the radius of a base, and h is the height.

EXAMPLE 1 *Finding the Surface Area of a Prism Using Theorem 12.2*

Find the surface area of the right prism.

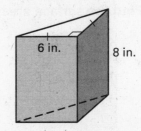

6 in.   8 in.

Reteaching with Practice

For use with pages 728–734

SOLUTION

Each base of the prism is a right triangle with base and height of 6 inches. Using the formula for the area of a triangle, the area of each base is $B = \frac{1}{2}(6)(6) = 18$ square inches.

To find the perimeter of each base, you need to find the length of the third side of the triangle. Because the triangle is an isosceles right triangle, it is a 45°-45°-90° triangle. So, the hypotenuse is $6\sqrt{2}$. The perimeter of each base is $P = 6 + 6 + 6\sqrt{2} = 12 + 6\sqrt{2}$. So, the surface area is

$$S = 2B + Ph = 2(18) + \left(12 + 6\sqrt{2}\right)(8) \approx 199.9 \text{ square inches.}$$

Exercises for Example 1

Find the surface area of the right prism.

1.
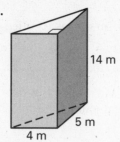
14 m
5 m
4 m

2.
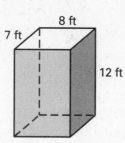
8 ft
7 ft
12 ft

3.
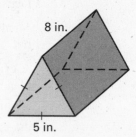
8 in.
5 in.

EXAMPLE 2 *Finding the Surface Area of a Cylinder*

Find the surface area of the right cylinder.

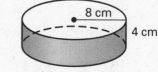
8 cm
4 cm

SOLUTION

The cylinder has an 8 cm radius and a 4 cm height.

$S = 2\pi r^2 + 2\pi rh$	Formula for surface area of a right cylinder
$= 2\pi(8)^2 + 2\pi(8)(4)$	Substitute.
$= 128\pi + 64\pi$	Simplify.
$= 192\pi$	Add.
≈ 603.2	Use a calculator.

The surface area is about 603.2 square centimeters.

Exercises for Example 2

Find the surface area of the right cylinder.

4.
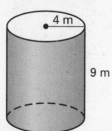
4 m
9 m

5.
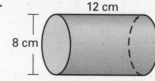
12 cm
8 cm

6.
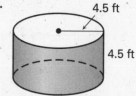
4.5 ft
4.5 ft

NAME _____ DATE _____

Quick Catch-Up for Absent Students

For use with pages 727–734

The items checked below were covered in class on (date missed) _____

Activity 12.2: Investigating Surface Area (p. 727)

____ **Goal:** Determine how a net can be used to find the surface area of a polyhedron.

Lesson 12.2: Surface Area of Prisms and Cylinders

____ **Goal 1:** Find the surface area of a prism. (pp. 728–729)

Material Covered:

____ Student Help: Study Tip

____ Example 1: Finding the Surface Area of a Prism

____ Student Help: Study Tip

____ Student Help: Look Back

____ Example 2: Using Theorem 12.2

Vocabulary:

prism, p. 728	bases, p. 728
lateral faces, p. 728	right prism, p. 728
oblique prism, p. 728	surface area, p. 728
lateral area, p. 728	net, p. 729

____ **Goal 2:** Find the surface area of a cylinder. (pp. 730–731)

Material Covered:

____ Example 3: Finding the Surface Area of a Cylinder

____ Example 4: Finding the Height of a Cylinder

Vocabulary:

cylinder, p. 730	right cylinder, p. 730
lateral area of a cylinder, p. 730	surface area of a cylinder, p. 730

____ Other (specify) _____

_____

Homework and Additional Learning Support

____ Textbook (specify) <u>pp. 731–734</u> _____

_____

____ Internet: Extra Examples at www.mcdougallittell.com

____ *Reteaching with Practice* worksheet (specify exercises)_____

____ *Personal Student Tutor* for Lesson 12.2

NAME _____ DATE _____

Real-Life Application:
When Will I Ever Use This?

For use with pages 728–734

Saint Bernard

The Saint Bernard, a people and family oriented breed of dog, originated in Switzerland in the 18th century where it was used primarily to guide monks over snow covered trails. It stands from 25 to 29 inches high and weighs anywhere from 140 to 170 pounds. Saint Bernards can have both rough and smooth coats that can be white with red markings, red with white markings, white with brindle, or brindle with white. They have a sturdy, muscular body with an imposing head and alert facial expressions.

Although undoubtedly cute, this breed of dog requires a lot of care and affection. These large dogs shed a lot of hair and drool constantly. They need to be fed twice a day to sustain them. Despite their size, they are a loving dog that will protect their family and home.

In Exercises 1–5, use the following information.

You are building a doghouse for your Saint Bernard, Hercules. You want the doghouse to match the color and style of your house as much as possible, so you plan to paint it and shingle the roof. The doghouse will be five feet long and four feet wide. The height of the doghouse is four feet and the roof peak rises two additional feet.

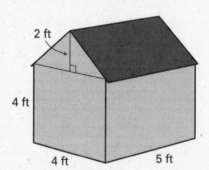

1. Find the amount of painted area (not including the area to be shingled).

2. If one gallon of paint covers 400 square feet, how many gallons are needed if you plan to put two coats outside and paint all the side walls of the inside twice?

3. Find the area of the roof that is to be shingled.

4. Shingles come in squares. One square of shingles covers 100 square feet. How many squares of shingles will you need to buy?

5. Will you have enough shingles left to re-shingle the storage shed if its roof has twice the area of the roof of the doghouse?

Geometry
Chapter 12 Resource Book

Challenge: Skills and Applications

For use with pages 728–734

1. Refer to the diagram.

   a. Sketch the solid that results after the net has been folded.

   b. Find the surface area of the solid.

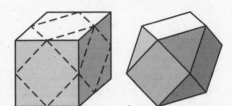

2. A *cuboctahedron* has 6 square faces and 8 equilateral triangle faces. It can be made by slicing off the corners of a cube, as shown.

   a. Sketch a net for a cuboctahedron.

   b. If each edge of a cuboctahedron has length 3 cm, find the surface area of the cuboctahedron.

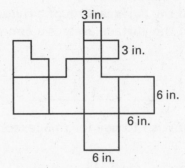

In Exercises 3–5, find the surface area of the oblique prism.

3.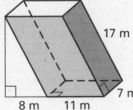
   4.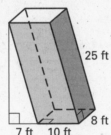
   5.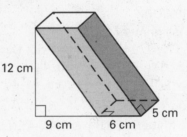

6. A right prism has a square base, a surface area of 512 in.², and a height of 12 in. Find the side length of the square base.

7. A right circular cylinder has a surface area of 180π mm² and a base of radius 6 mm. Find the height of the cylinder.

8. A right circular cylinder has a surface area of 168π ft² and a height of 5 ft. Find the radius of the cylinder.

9. The diagram at the right shows a net for a cereal box. A small spider is at position *S*, and a bug is at position *B*.

   a. According to the net shown, what is the apparent distance between the spider and the bug?

   b. Sketch a different net for the same cereal box, showing a shorter path from the spider to the bug. What is the shortest possible distance that the spider would have to walk to get to the bug?

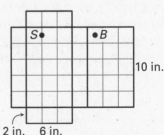

TEACHER'S NAME _____ CLASS _____ ROOM _____ DATE _____

Lesson Plan

2-day lesson (See *Pacing the Chapter*, TE pages 716C–716D)   **For use with pages 735–742**

GOALS 1. Find the surface area of a pyramid.
2. Find the surface area of a cone.

State/Local Objectives _____

_____

✓ **Check the items you wish to use for this lesson.**

STARTING OPTIONS

____ Homework Check: TE page 731: Answer Transparencies
____ Warm-Up or Daily Homework Quiz: TE pages 735 and 734, CRB page 36, or Transparencies

TEACHING OPTIONS

____ Motivating the Lesson: TE page 736
____ Lesson Opener (Application): CRB page 37 or Transparencies
____ Technology Activity with Keystrokes: CRB pages 38–40
____ Examples: Day 1: 1–3, SE pages 735–737; Day 2: See the Extra Examples.
____ Extra Examples: Day 1 or Day 2: 1–3, TE pages 736–737 or Transp.
____ Closure Question: TE page 737
____ Guided Practice: SE page 738   Day 1: Exs. 1–13; Day 2: See Checkpoint Exs. TE pages 736–737

APPLY/HOMEWORK
Homework Assignment

____ Basic   Day 1: 14–38 even, 44–46; Day 2: 15–39 odd, 43, 50–53; Quiz 1: 1–6
____ Average   Day 1: 14–38 even, 44–46; Day 2: 15–43 odd, 50–53; Quiz 1: 1–6
____ Advanced   Day 1: 14–38 even, 44–46; Day 2: 15–43 odd, 50–53; Quiz 1: 1–6

Reteaching the Lesson

____ Practice Masters: CRB pages 41–43 (Level A, Level B, Level C)
____ Reteaching with Practice: CRB pages 44–45 or Practice Workbook with Examples
____ Personal Student Tutor

Extending the Lesson

____ Applications (Interdisciplinary): CRB page 47
____ Math & History: SE page 742; CRB page 48; Internet
____ Challenge: SE page 741; CRB page 49 or Internet

ASSESSMENT OPTIONS

____ Checkpoint Exercises: Day 1 or Day 2: TE pages 736–737 or Transp.
____ Daily Homework Quiz (12.3): TE page 741, CRB page 53, or Transparencies
____ Standardized Test Practice: SE page 741; TE page 741; STP Workbook; Transparencies
____ Quiz (12.1–12.3): SE page 742; CRB page 50

Notes _____

_____

_____

Lesson 12.3

TEACHER'S NAME _____ CLASS _____ ROOM _____ DATE _____

Lesson Plan for Block Scheduling

1-day lesson (See *Pacing the Chapter,* TE pages 716C–716D)   **For use with pages 735–742**

GOALS   1. **Find the surface area of a pyramid.**
2. **Find the surface area of a cone.**

State/Local Objectives _____

_____

_____

✓ **Check the items you wish to use for this lesson.**

CHAPTER PACING GUIDE	
Day	**Lesson**
1	Assess Ch. 11; 12.1 (all)
2	12.2 (all); **12.3 (begin)**
3	**12.3 (end)**; 12.4 (begin)
4	12.4 (end); 12.5 (begin)
5	12.5 (end); 12.6 (all)
6	12.7 (all)
7	Review Ch. 12; Assess Ch. 12

STARTING OPTIONS
____ Homework Check: TE page 731: Answer Transparencies
____ Warm-Up or Daily Homework Quiz: TE pages 735 and
        734, CRB page 36, or Transparencies

TEACHING OPTIONS
____ Motivating the Lesson: TE page 736
____ Lesson Opener (Application): CRB page 37 or Transparencies
____ Technology Activity with Keystrokes: CRB pages 38–40
____ Examples: Day 2: 1–3, SE pages 735–737; Day 3: See the Extra Examples.
____ Extra Examples: Day 2 or Day 3: 1–3, TE pages 736–737 or Transp.
____ Closure Question: TE page 737
____ Guided Practice: SE page 738   Day 2: Exs. 1–13; Day 3: See Checkpoint Exs. TE pages 736–737

APPLY/HOMEWORK
Homework Assignment (See also the assignments for Lessons 12.2 and 12.4.)
____ Block Schedule: Day 2: 14–38 even, 42, 44–46; Day 3: 15–43 odd, 43, 50–53; Quiz 1: 1–6

Reteaching the Lesson
____ Practice Masters: CRB pages 41–43 (Level A, Level B, Level C)
____ Reteaching with Practice: CRB pages 44–45 or Practice Workbook with Examples
____ Personal Student Tutor

Extending the Lesson
____ Applications (Interdisciplinary): CRB page 47
____ Math & History: SE page 742; CRB page 48; Internet
____ Challenge: SE page 741; CRB page 49 or Internet

ASSESSMENT OPTIONS
____ Checkpoint Exercises: Day 2 or Day 3: TE pages 736–737 or Transp.
____ Daily Homework Quiz (12.3): TE page 741, CRB page 53, or Transparencies
____ Standardized Test Practice: SE page 741; TE page 741; STP Workbook; Transparencies
____ Quiz (12.1–12.3): SE page 742; CRB page 50

Notes _____

_____

_____

Lesson 12.3

WARM-UP EXERCISES

For use before Lesson 12.3, pages 735–742

Find the surface area of the solid.

1. right rectangular prism: length = 2 cm, width = 3 cm, height = 10 cm

2. right cylinder: height = 12 ft, radius = 5 ft

3. right pentagonal prism: area of a base = 62 m², perimeter of a base = 30 m, height = 5 m

4. right cylinder: height = 6 in., diameter = 3 in.

DAILY HOMEWORK QUIZ

For use after Lesson 12.2, pages 727–734

1. How many lateral faces and how many lateral edges does an oblique octagonal prism have?

2. Find the surface area of the right prism. Round your result to two decimal places.

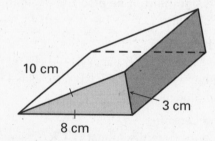

10 cm   3 cm   8 cm

3. Find the surface area of a right cylinder that has a base diameter of 8 inches and a height of 15 inches. Round your result to two decimal places.

4. Solve for the variable given that the surface area of the right prism is 208 m².

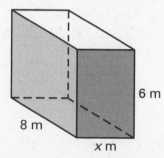

6 m   8 m   x m

NAME _____ DATE _____

Application Lesson Opener

For use with pages 735–742

The Sioux tipi is representative of the 3-pole tipis lived in by many Native Americans. An average family lodge uses 17 25-foot poles: 15 for the frame and 2 for the smoke flaps. The tipi is tilted, steeper at the back and more gently sloping at the front for the smoke hole. The inside measures about 20 feet from rear to door and 17.5 feet across.

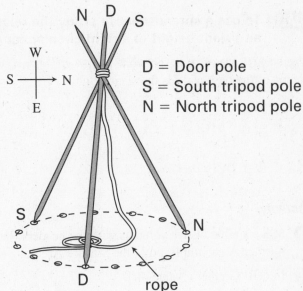

D = Door pole
S = South tripod pole
N = North tripod pole

The diagram at the right shows a design for a 15-pole frame, in which three poles are erected first to form a tripod.

The diagram below shows the cover of a tipi. It has a Cheyenne design with symbols of the sun, moon, rain, and a buffalo. The cover is stretched around a frame of poles and pegged to the ground.

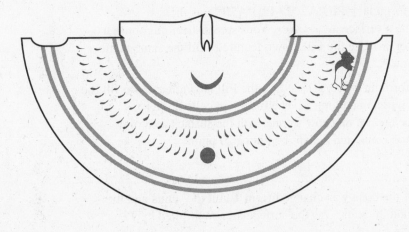

1. Sketch a tipi. Is it a pyramid or a cone? Explain your answer.

NAME _____   DATE _____

Technology Activity

For use with pages 735–742

GOAL **To use a spreadsheet to study the relationship between the radius and slant height of a right circular cone and the cone's surface area**

The surface area S of a right cone is $S = \pi r^2 + \pi r l$, where r is the radius of the base and l is the slant height.

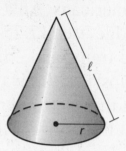

Activity

1 Make a table of five columns with the headers Radius, Height, Slant Height, Surface Area, and Ratio of the Surface Areas.

2 In cell A2 enter a value of 1.

3 In cell A3 enter the formula =A2+1 and use the Fill Down feature to create values up to 15.

4 In cell C2, enter a value of 20.

5 In column C use the Fill Down feature to fill the same number of rows as in column A.

6 In cell B2 use the formula =SQRT(C2^2−A2^2) as the formula for the height of the cone. Use the Fill Down feature to fill the same number of rows as in column A.

7 In cell D2 use the formula =PI()*A2^2+PI()*A2*C2 as the formula for the cone's surface area. (*Note:* Your spreadsheet might use a different formula for π.) Use the Fill Down feature to fill the same number of rows as in column A.

8 In cell E3 use the formula =D3/$D$2. Use the Fill Down feature to fill the same number of rows as in column A. The results in cells E3–E16 compare the surface area of that row's cone with the surface area of a cone with a radius of 1 unit and a slant height of 20 units.

Exercises

1. When the radius of the cone was changed from 1 unit to 7 units and the slant height remained 20 units, was the surface area multiplied by 7?

2. Create a spreadsheet like the one in the activity, but keep the radius constant at 1 unit and let the slant height vary from 20 units to 34 units, 1 unit at a time. Which dimension has a greater affect on the surface area for each unit of change: radius or slant height? Explain.

NAME _____ DATE _____

Technology Activity

For use with pages 735–742

EXCEL

1. Select cell A1.

   Radius `TAB` Height `TAB` Slant Height `TAB` Surface Area `TAB` Ratio of the Surface Areas `ENTER`

2. Select cell A2.

   1 `ENTER`

3. Select cell A3.

   =A2+1 `ENTER`

   Select cells A3–A16. From the **Edit** menu, choose **Fill Down**.

4. Select cell C2.

   20 `ENTER`

5. Select cells C2–C16. From the **Edit** menu, choose **Fill Down**.

6. Select cell B2.

   =SQRT(C2^2 − A2^2) `ENTER`

   Select cells B2–B16. From the **Edit** menu, choose **Fill Down**.

7. Select cell D2.

   =PI()*A2^2+PI()*A2*C2 `ENTER`

   Select cells D2–D16. From the **Edit** menu, choose **Fill Down**.

8. Select cell E3.

   =D3/$D$2 `ENTER`

   Select cells E3–E16. From the **Edit** menu, choose **Fill Down**.

   The results in cells E3–E16 compare the surface area of the cone in that row to the surface area of a cone with a radius of 1 unit and a slant height of 20 units.

Technology Activity

For use with pages 735–742

Keystrokes for Exercise 2 on page 38 of Chapter 12 Resource Book

EXCEL

Adjust the keystrokes given on page 39 as follows.

To keep the radius (cells A2–A16) constant at 1 unit:

> Select cell A2.
>
> 1 **ENTER**
>
> Select cells A2–A16. From the **Edit** menu, choose **Fill Down**.

To vary the slant height (cells C3–C16):

> Select cell C2.
>
> 20 **ENTER**
>
> Select cell C3.
>
> =C2+1 **ENTER**
>
> Select cells C3–C16. From the **Edit** menu, choose **Fill Down**.

The height of the cone (cells B2–B16) will still be =SQRT(C2^2−A2^2)

Note: Check column E. You may need to change the formula "=D3/$D$2" to match the row numbers where you have positioned your spreadsheet for Exercise 2.

Lesson 12.3

NAME _____ DATE _____

Practice A
For use with pages 735–742

Use the diagram at the right.

1. Give the mathematical name of the solid.

2. What kind of figure is the base?

3. What kind of figure is each lateral face?

4. How many lateral faces does the solid have?

5. Name three lateral edges.

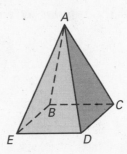

Find the slant height *l* of the regular pyramid or cone.

6.

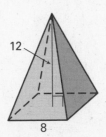 

7.

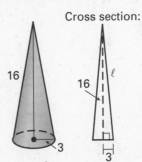

Find the surface area of the regular pyramid.

8.

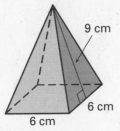

9.

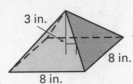

10.

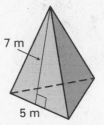

Find the surface area of the right cone. Leave your answers in terms of π.

11.

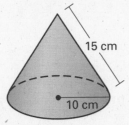

12.

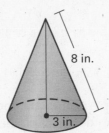

13.

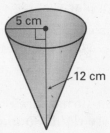

Sketch the described solid and find its surface area. Round the result to one decimal place.

14. A regular pyramid has a square base with a base edge of 10 centimeters and a height of 6 centimeters.

NAME _____ DATE _____

Practice B

For use with pages 735–742

Find the slant height *l* of the regular pyramid or cone.

1.
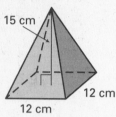
15 cm
12 cm
12 cm

2.
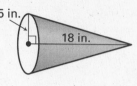
5 in.
18 in.

3.
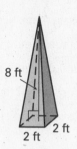
8 ft
2 ft
2 ft

Find the surface area of the regular pyramid.

4.
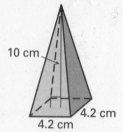
10 cm
4.2 cm
4.2 cm

5.
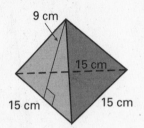
9 cm
15 cm
15 cm
15 cm

6.
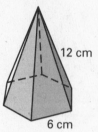
12 cm
6 cm

Find the surface area of the right cone. Leave your answers in terms of π.

7.
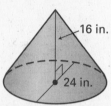
16 in.
24 in.

8.
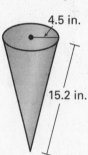
4.5 in.
15.2 in.

9.
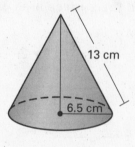
13 cm
6.5 cm

Sketch the described solid and find its surface area. Round the result to one decimal place.

10. A regular pyramid has a square base with a base edge of 10 centimeters and a height of 6 centimeters.

11. A right cone has a diameter of 14 centimeters and a slant height of 11 centimeters.

Find the surface area of the solid. The pyramids are regular and the prisms, cylinders, and cones are right. Round the result to one decimal place.

12.
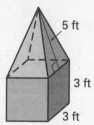
5 ft
3 ft
3 ft

13.
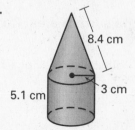
8.4 cm
5.1 cm
3 cm

14.
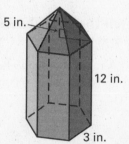
5 in.
12 in.
3 in.

NAME _____ DATE _____

Practice C

For use with pages 735–742

Find the slant height *l* of the regular pyramid or cone.

1.

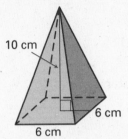

10 cm

6 cm

6 cm

2.

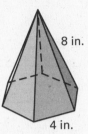

8 in.

4 in.

3.

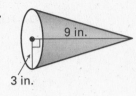

9 in.

3 in.

Find the surface area of the regular pyramid.

4.

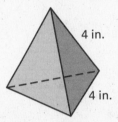

4 in.

4 in.

5.

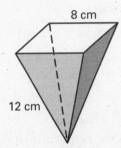

8 cm

12 cm

6.

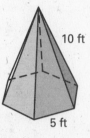

10 ft

5 ft

Find the surface area of the right cone. Leave your answers in terms of π.

7.

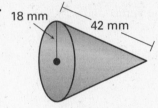

18 mm

42 mm

8.

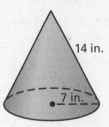

14 in.

7 in.

9.

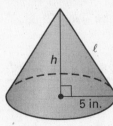

6 cm

8.4 cm

Find the missing measurements of the regular pyramid or right cone.

10. $P = 56$ cm

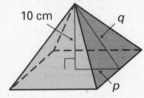

10 cm

q

p

11. $S = 219.9$ in.²

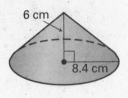

h

ℓ

5 in.

12. $S = 273.53$ cm²,
$P = 36$ cm

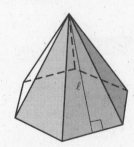

ℓ

Reteaching with Practice

For use with pages 735–742

GOAL Find the surface area of a pyramid and find the surface area of a cone

---

VOCABULARY

A **pyramid** is a polyhedron in which the base is a polygon and the lateral faces are triangles with a common vertex.

The intersection of two lateral faces of a pyramid is called a **lateral edge**.

The intersection of the base of a pyramid and a lateral face is called a **base edge**.

The **altitude**, or **height**, of a pyramid is the perpendicular distance between the base and the vertex.

A **regular pyramid** has a regular polygon for a base and its height meets the base at its center.

The **slant height** of a regular pyramid is the altitude of any lateral face.

A **circular cone**, or **cone**, has a circular base and a vertex that is not in the same plane as the base.

The **altitude**, or **height**, is the perpendicular distance between the vertex and the base.

In a **right cone**, the height meets the base at its center.

In a right cone, the **slant height** is the distance between the vertex and a point on the base edge.

The **lateral surface** of a cone consists of all segments that connect the vertex with points on the base edge.

Theorem 12.4   Surface Area of a Regular Pyramid
The surface area S of a regular pyramid is $S = B + \frac{1}{2}P\ell$, where B is the area of the base, P is the perimeter of the base, and ℓ is the slant height.

Theorem 12.5 Surface Area of a Right Cone
The surface area S of a right cone is $S = \pi r^2 + \pi r \ell$, where r is the radius of the base and ℓ is the slant height.

EXAMPLE 1   *Finding the Surface Area of a Pyramid*

Find the surface area of the regular pyramid shown.

SOLUTION

To find the surface area of the regular pyramid, start by finding the area of the base.

Because the base is a square, the area is s^2. So, the area of the base is 5^2, or 25 square feet.

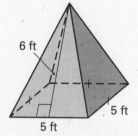

6 ft

5 ft

5 ft

Lesson 12.3

NAME _____ DATE _____

Reteaching with Practice

For use with pages 735–742

Now you can find the surface area, using 25 for the area of the base, B.

$S = B + \frac{1}{2}P\ell$ Write formula.

$= 25 + \frac{1}{2}(4 \cdot 5)(6)$ Substitute.

$= 85$ Simplify.

So, the surface area is 85 square feet.

Exercises for Example 1

Find the surface area of the regular pyramid.

1.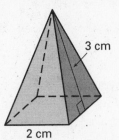

2.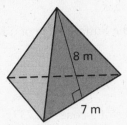

3.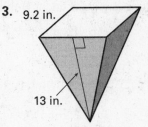

EXAMPLE 2   *Finding the Surface Area of a Cone*

Find the surface area of the right cone shown.

SOLUTION

With a radius of 3 meters and a slant height of 7 meters given, use the formula to find the surface area.

$S = \pi r^2 + \pi r \ell$ Write formula.

$= \pi(3)^2 + \pi(3)(7)$ Substitute.

$= 9\pi + 21\pi$ Simplify.

$= 30\pi$ Simplify.

So, the surface area of the cone is 30π square meters, or about 94.2 square meters.

Exercises for Example 2

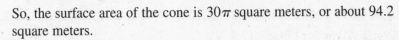

Find the surface area of the right cone.

4.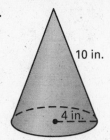

5.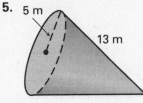

6.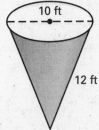

NAME _____ DATE_____

Quick Catch-Up for Absent Students

For use with pages 735–742

The items checked below were covered in class on (date missed) _____

Lesson 12.3: Surface Area of Pyramids and Cones

_____ **Goal 1:** Find the surface area of a pyramid. (pp. 735–736)

Material Covered:

_____ Student Help: Study Tip

_____ Student Help: Study Tip

_____ Example 1: Finding the Area of a Lateral Face

_____ Student Help: Look Back

_____ Example 2: Finding the Surface Area of a Pyramid

Vocabulary:

    pyramid, p. 735   regular pyramid, p. 735

_____ **Goal 2:** Find the surface area of a cone. (p. 737)

Material Covered:

_____ Student Help: Study Tip

_____ Example 3: Finding the Surface Area of a Right Cone

Vocabulary:

    circular cone, p. 737   cone, p. 737

    right cone, p. 737   lateral surface, p. 737

_____ Other (specify) _____

_____

Homework and Additional Learning Support

_____ Textbook (specify) pp. 738–742 _____

_____

_____ *Reteaching with Practice* worksheet (specify exercises)_____

_____ *Personal Student Tutor* for Lesson 12.3

Lesson 12.3

Interdisciplinary Application

For use with pages 735–742

Aztecs

HISTORY The Aztecs of Pre-Columbian Central America were in many respects an advanced culture. Mexico City at the time of the Spanish conquest had a population of one million. This was larger than Madrid. Their advanced agricultural techniques allowed them to support large numbers of people in cities.

Like the Egyptians, the Aztecs built pyramids. One difference between the Aztec and Egyptian pyramids was in shape. The lateral faces of Egyptian pyramids were triangular while their Aztec counterparts were trapezoidal. The Aztecs built their pyramids with a platform at the top with a room on the platform. The room was used by the Aztecs for ceremonies. This was the second difference since the Egyptians used their pyramids as monuments for the Pharaohs.

In Exercises 1–5, use the following information.

Suppose a regular Aztec pyramid has a square base with a 150 foot base edge. It is 87 feet tall and the angle of the slant is 55°. To find the surface area, find the area of a pyramid having triangular sides with the same dimensions and subtract the surface area of the missing top.

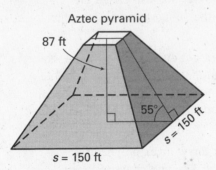
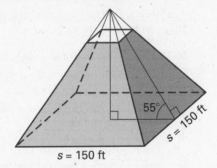

1. Find the slant height of the large triangular pyramid with the given measurements. Round your result to the nearest foot.

2. Find the surface area of the triangular pyramid.

3. Use similar solids to find (a) the length of the base of the small triangular pyramid, and (b) the height of the small triangular pyramid. Round your results to the nearest foot.

4. Find the surface area of (a) the small pyramid, and (b) the Aztec trapezoidal pyramid.

NAME _____ DATE _____

Math and History Application

For use with page 742

MATH A *regular polyhedron* is a solid whose faces are all regular polygons. That is, in each face, the sides all have the same length and the interior angles all have the same measure. For example, a *cube* has six congruent faces, each one a square. A *tetrahedron* has four congruent faces, each one an equilateral triangle.

There are five regular polyhedra: the cube, tetrahedron, octahedron (8 faces), dodecahedron (12 faces), and icosahedron (20 faces).

  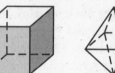

Regular tetrahedron
4 faces, 4 vertices, 6 edges

Cube
6 faces, 8 vertices, 12 edges

Regular octahedron
8 faces, 6 vertices, 12 edges

Regular dodecahedron
12 faces, 20 vertices, 30 edges

Regular icosahedron
20 faces, 12 vertices, 30 edges

The five regular polyhedra shown above are also called the *Platonic solids*, after the Greek mathematician and philosopher Plato (427 B.C.–347 B.C.).

HISTORY The Platonic solids were not discovered by Plato, however. The tetrahedron, essentially a pyramid, and octahedron, a double pyramid with a square base, was known to the ancient Egyptians, while the bronze castings of dodecahedra were found from even earlier periods.

1. Euler's theorem states that, for any convex polyhedron, the number of faces (F), vertices (V), and edges (E) are related by the formula $F + V = E + 2$. Verify Euler's formula for the cube and tetrahedron.

2. The area A of an equilateral triangle is given by the formula $A = \dfrac{\sqrt{3}}{4}s^2$, where s is the length of the side. Use this formula to find the surface area of an icosahedron of side length s.

NAME _____ DATE _____

Challenge: Skills and Applications

For use with pages 735–742

In Exercises 1–5, give answers as decimals rounded to the nearest hundredth, where appropriate.

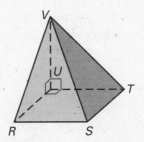

1. In the diagram, *RSTU* is a square, *TU* = 21 in., *UV* = 20 in., and $\overline{UV}$ is perpendicular to the plane containing *RSTU*.

   a. Find *TV* and *SV*.

   b. Sketch a net of the pyramid.

   c. Find the surface area of the pyramid.

2. The base of a regular pyramid is a square whose sides have length 8 cm. If a lateral edge of this pyramid has length 20 cm, what is the surface area of the pyramid?

3. The base of a regular pyramid is a hexagon whose sides have length 5 ft. If a lateral edge of this pyramid has length 9 ft, what is the surface area of the pyramid?

4. The sector shown in the diagram can be rolled up to form the lateral surface of a cone whose lateral surface area is 26 cm².

   a. Find the slant height of the cone.

   b. Find the height and radius of the cone.

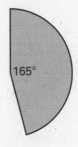

165°

5. The cone shown in the diagram has a base radius of 12 in. and a height of 35 in. The lateral surface of the cone shown in the diagram can be unrolled to form a sector like the one shown in Exercise 4.

   a. Find the central angle and the radius of the sector.

   b. Find the surface area of the cone.

NAME _____ DATE _____

Quiz 1

For use after Lessons 12.1–12.3

State whether the polyhedron is regular and/or convex. Then calculate the number of vertices of the solid using the given information. *(Lesson 12.1)*

1. 5 faces; 2 triangles and 3 rectangles

2. 4 faces; all triangles

3. 6 faces; 2 squares and 4 rectangles

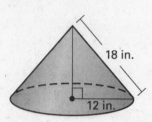

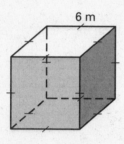

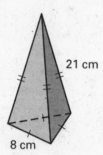

Find the surface area of the solid. Round your result to two decimal places. *(Lessons 12.2 and 12.3)*

4.

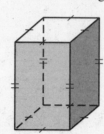

18 in.

12 in.

5.

6 m

6.

21 cm

8 cm

TEACHER'S NAME _____ CLASS _____ ROOM _____ DATE _____

Lesson Plan

2-day lesson (See *Pacing the Chapter*, TE pages 716C–716D)   For use with pages 743–750

GOALS
1. **Use volume postulates.**
2. **Find the volume of prisms and cylinders in real-life.**

State/Local Objectives _____

_____

✓ **Check the items you wish to use for this lesson.**

STARTING OPTIONS
____ Homework Check: TE page 738: Answer Transparencies
____ Warm-Up or Daily Homework Quiz: TE pages 743 and 741, CRB page 53, or Transparencies

TEACHING OPTIONS
____ Motivating the Lesson: TE page 744
____ Lesson Opener (Application): CRB page 54 or Transparencies
____ Technology Activity with Keystrokes: CRB pages 55–56
____ Examples: Day 1: 1–3, SE pages 743–745; Day 2: SE page 745
____ Extra Examples: Day 1: TE pages 744–745 or Transp.; Day 2: TE page 745 or Transp.
____ Technology Activity: SE page 750
____ Closure Question: TE page 745
____ Guided Practice: SE page 746   Day 1: Exs. 1–6; Day 2: Exs. 7–9

APPLY/HOMEWORK
Homework Assignment
____ Basic   Day 1: 10–33; Day 2: 34–49
____ Average   Day 1: 10–33; Day 2: 34–49
____ Advanced   Day 1: 10–33; Day 2: 34–50

Reteaching the Lesson
____ Practice Masters: CRB pages 57–59 (Level A, Level B, Level C)
____ Reteaching with Practice: CRB pages 60–61 or Practice Workbook with Examples
____ Personal Student Tutor

Extending the Lesson
____ Applications (Real-Life): CRB page 63
____ Challenge: SE page 749; CRB page 64 or Internet

ASSESSMENT OPTIONS
____ Checkpoint Exercises: Day 1: TE pages 744–745 or Transp.; Day 2: TE page 745 or Transp.
____ Daily Homework Quiz (12.4): TE page 749, CRB page 67, or Transparencies
____ Standardized Test Practice: SE page 749; TE page 749; STP Workbook; Transparencies

Notes _____

_____

_____

Lesson Plan for Block Scheduling

1-day lesson (See *Pacing the Chapter,* TE pages 716C–716D)   For use with pages 743–750

GOALS   1. Use volume postulates.
2. Find the volume of prisms and cylinders in real-life.

State/Local Objectives _____

_____

_____

CHAPTER PACING GUIDE	
Day	**Lesson**
1	Assess Ch. 11; 12.1 (all)
2	12.2 (all); 12.3 (begin)
3	12.3 (end); **12.4 (begin)**
4	**12.4 (end)**; 12.5 (begin)
5	12.5 (end); 12.6 (all)
6	12.7 (all)
7	Review Ch. 12; Assess Ch. 12

✓ **Check the items you wish to use for this lesson.**

STARTING OPTIONS
____ Homework Check: TE page 738: Answer Transparencies
____ Warm-Up or Daily Homework Quiz: TE pages 743 and
        741, CRB page 53, or Transparencies

TEACHING OPTIONS
____ Motivating the Lesson: TE page 744
____ Lesson Opener (Application): CRB page 54 or Transparencies
____ Technology Activity with Keystrokes: CRB pages 55–56
____ Examples: Day 3: 1–3, SE pages 743–745; Day 4: SE page 745
____ Extra Examples: Day 3: TE pages 744–745 or Transp.; Day 4: TE page 745 or Transp.
____ Technology Activity: SE page 750
____ Closure Question: TE page 745
____ Guided Practice: SE page 746   Day 3: Exs. 1–6; Day 4: Exs. 7–9

APPLY/HOMEWORK
Homework Assignment (See also the assignments for Lessons 12.3 and 12.5.)
____ Block Schedule: Day 3: 10–33; Day 4: 34–49

Reteaching the Lesson
____ Practice Masters: CRB pages 57–59 (Level A, Level B, Level C)
____ Reteaching with Practice: CRB pages 60–61 or Practice Workbook with Examples
____ Personal Student Tutor

Extending the Lesson
____ Applications (Real-Life): CRB page 63
____ Challenge: SE page 749; CRB page 64 or Internet

ASSESSMENT OPTIONS
____ Checkpoint Exercises: Day 3: TE pages 744–745 or Transp.; Day 4: TE page 745 or Transp.
____ Daily Homework Quiz (12.4): TE page 749, CRB page 67, or Transparencies
____ Standardized Test Practice: SE page 749; TE page 749; STP Workbook; Transparencies

Notes _____

_____

_____

Find the area of each figure.

1. a square with sides of length 4

2. a circle with a radius of 7 cm

3. an equilateral triangle with sides of length 4 meters

4. a regular hexagon with sides of length 2 feet

DAILY HOMEWORK QUIZ

For use after Lesson 12.3, pages 735–742

1. Find the surface area of the
   regular pyramid.

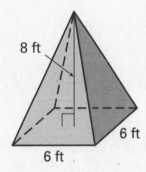

8 ft

6 ft

6 ft

2. Find the slant height of a right cone that has a height
   of 15 mm and a radius of 9 mm.

3. Find the surface area of a right cone that has a height
   of 12 in. and a radius of 9 in. Leave your answer in
   terms of π.

4. Find the missing measurements
   of the regular triangular pyramid
   if $S = 225$ m^2 and $P = 30$ m.

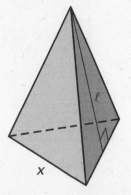

ℓ

x

Application Lesson Opener

For use with pages 743–749

Grocery stores are filled with many sizes and shapes of prisms (boxes) and cylinders (cans) filled with food products. The size and shape of a container determines its *volume* (the number of cubic units contained in its interior).

1. A type of cereal is available in the three different boxes shown. Find the volume *V* of each box in cubic centimeters, using the formula *V* = length • width • height. Are the net weights proportional to the volumes? Should they be? Explain.

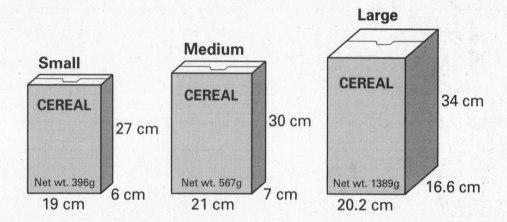

2. A type of chili is available in the three different cans shown. Find the volume *V* of each can in cubic centimeters, using the formula $V = \pi \cdot (\text{radius})^2 \cdot \text{height}$. Are the net weights proportional to the volumes? Should they be? Explain.

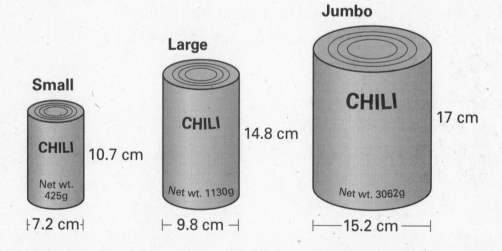

Geometry
Chapter 12 Resource Book

NAME _____ DATE _____

Technology Activity Keystrokes

For use with page 750

EXCEL

Select cell A1.

Volume V [TAB] Radius r [TAB] Height h = V/(Pi r^2) [TAB] Surface area

SA=2Pi r^2+2Pirh [ENTER]

Select cell A2.

72 [TAB] 2 [TAB] =A2/(PI()*B2^2) [TAB]

=2*PI()*B2^2+2*PI()*B2*C2 [ENTER]

Select cell A3.

=A2 [TAB] =B2+0.05 [TAB] =A3/(PI()*B3^2) [TAB]

=2*PI()*B3^2+2*PI()*B3*C3 [ENTER]

Select cells A, B, C, and D in Rows 3–30. From the **Edit** menu, choose **Fill Down**.

NAME _____ DATE _____

Practice A

For use with pages 743–749

Find the number of unit cubes that will fit in the box. Explain your reasoning.

1.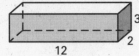
12   3   2

2.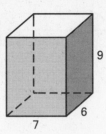
9   7   6

3.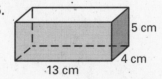
6   8   8

Find the volume of the right prism.

4.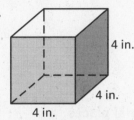
4 in.   4 in.   4 in.

5.
5 cm   4 cm   13 cm

6.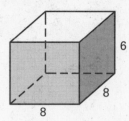
6 ft   4 ft   1 ft

7.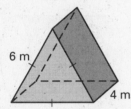
6 m   4 m

8.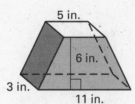
5 in.   6 in.   3 in.   11 in.

9.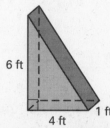
8 in.   4 in.

Find the volume of the right cylinder. Round the result to two decimal places.

10.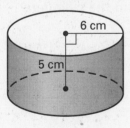
6 cm   5 cm

11.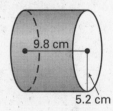
4 in.   11 in.

12.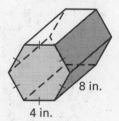
9.8 cm   5.2 cm

13. *Swimming Pool* A swimming pool measures 40 feet long by 20 feet wide. The pool is filled to a depth of 6 feet. Find the volume of the water in the pool.

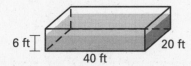
6 ft   40 ft   20 ft

14. *Swimming Pool* A common design for swimming pools is for the depth to change gradually from the shallow end to the deep end. Use the dimensions shown to find the volume of water the pool can hold.

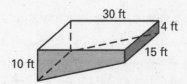
30 ft   4 ft   15 ft   10 ft

Geometry
Chapter 12 Resource Book

57

NAME _____ DATE _____

Practice B

Find the volume of the right prism.

1.

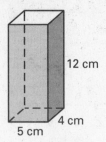

12 cm

5 cm 4 cm

2.

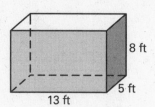

8 ft

5 ft

13 ft

3.

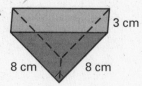

3 cm

8 cm 8 cm

4.

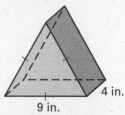

4 in.

9 in.

5.

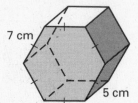

7 cm

5 cm

6.

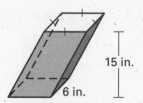

15 in.

6 in.

Find the volume of the right cylinder. Round the result to two decimal places.

7.

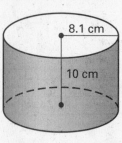

8.1 cm

10 cm

8.

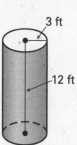

3 ft

12 ft

9.

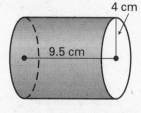

4 cm

9.5 cm

Solve for the variable using the given measurements. The prisms and the cylinders are right.

10. Volume = 525 cm³

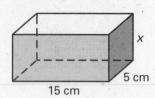

x

5 cm

15 cm

11. Volume = 2420 ft³

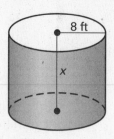

8 ft

x

12. Volume = 455 in.³

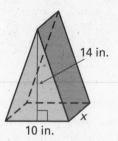

14 in.

10 in.

x

In Exercises 13 and 14, make a sketch of the solid and find its volume. Round the result to two decimal places.

13. A prism has a square base with 5 foot sides and a height of 2.5 feet.

14. A cylinder has a diameter of 23 inches and a height of 16 inches.

15. *Pillars* How much plaster of paris is needed to make four miniature pillars for a model of a home if the pillars are regular hexagonal prisms with a height of 12 in. and base edges of length 2 in.?

Base of pillar

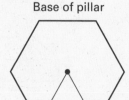

Ex. 15

2 in.

Lesson 12.4

NAME _____ DATE _____

Practice C

For use with pages 743–749

Find the volume of the prism.

1.

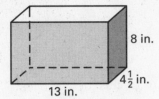

8 in.

13 in. $4\frac{1}{2}$ in.

2.

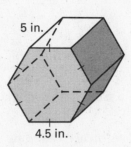

5 in.

4.5 in.

3.

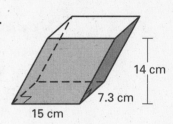

14 cm

7.3 cm

15 cm

Find the volume of the cylinder.

4.

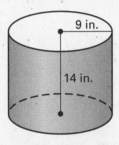

9 in.

14 in.

5.

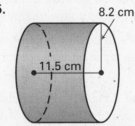

8.2 cm

11.5 cm

6.

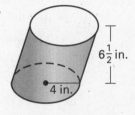

$6\frac{1}{2}$ in.

4 in.

7. Find the volume of a cube with 6 cm edges.

8. Find the volume of a rectangular prism that is 5 in. by 6 in. by 3 in.

9. Find the volume of a cylinder with a 2 m radius and a 4 m height.

10. Find the volume of a cylinder with a base area of 625π in.^{2} and a height of 25 in.

Solve for the variable using the given measurements. The prisms and the cylinders are right.

11. Volume = 200 cm^3

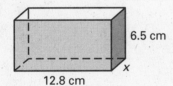

6.5 cm

12.8 cm

x

12. Volume = 475 ft^3

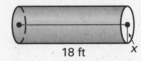

18 ft x

13. Volume = 665 in.^{3}

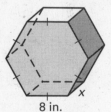

x

8 in.

Draw the prism or cylinder formed by the net. Then find its volume.

14.

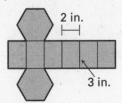

2 in.

3 in.

15.

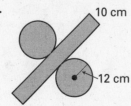

10 cm

12 cm

16.

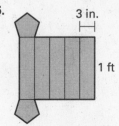

3 in.

1 ft

NAME _____ DATE _____

Reteaching with Practice

For use with pages 743–749

GOAL **Use volume postulates and find the volume of prisms and cylinders**

VOCABULARY

The **volume of a solid** is the number of cubic units contained in its interior. Volume is measured in cubic units.

Postulate 27 Volume of a Cube   The volume of a cube is the cube of the length of its side, or $V = s^3$.

Postulate 28 Volume Congruence Postulate   If two polyhedra are congruent, then they have the same volume.

Postulate 29 Volume Addition Postulate   The volume of a solid is the sum of the volumes of all its nonoverlapping parts.

Theorem 12.6 Cavalieri's Principle   If two solids have the same height and the same cross-sectional area at every level, then they have the same volume.

Theorem 12.7 Volume of a Prism   The volume V of a prism is $V = Bh$, where B is the area of a base and h is the height.

Theorem 12.8 Volume of a Cylinder   The volume V of a cylinder is $V = Bh = \pi r^2 h$, where B is the area of a base, h is the height, and r is the radius of a base.

EXAMPLE 1 *Finding Volumes*

Find the volume of the right cylinder and the right prism.

a.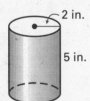

b.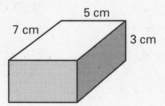

SOLUTION

a. The area B of the base is $\pi \cdot 2^2$, or 4π in.^{2}. Use $h = 5$ to find the volume.

$$V = Bh = 4\pi(5) = 20\pi \approx 62.83 \text{ in.}^3$$

b. The area B of the base is $(7)(5)$, or 35 cm^2. Use $h = 3$ to find the volume.

$$V = Bh = (35)(3) = 105 \text{ cm}^3$$

NAME _____ DATE _____

Reteaching with Practice

For use with pages 743–749

Exercises for Example 1

Find the volume of the right prism or the right cylinder.

1.

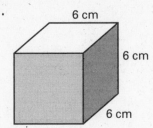

6 cm

6 cm

6 cm

2.

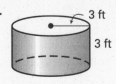

3 ft

3 ft

3.

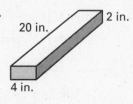

2 in.

20 in.

4 in.

EXAMPLE 2 *Using Volumes*

Use the measurements given to solve for *x*.

$V = 60$ cm^3

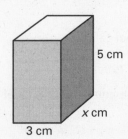

5 cm

x cm

3 cm

SOLUTION

The area of the base is $3x$ square centimeters.

$V = Bh$	Formula for volume of a right prism
$60 = (3x)(5)$	Substitute.
$60 = 15x$	Rewrite.
$\dfrac{60}{15} = x$	Divide each side by 15.
$4 = x$	Simplify.

Exercises for Example 2

Use the measurements given to solve for *x*.

4. $V = 283$ m^3

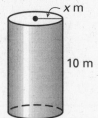

x m

10 m

5. $V = 64$ in.^{3}

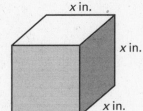

x in.

x in.

x in.

6. $V = 300$ ft^3

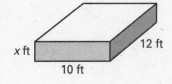

12 ft

x ft

10 ft

NAME _____ DATE _____

Quick Catch-Up for Absent Students

For use with pages 743–750

The items checked below were covered in class on (date missed) _____

Lesson 12.4: Volume of Prisms and Cylinders

____ **Goal 1:** Use volume postulates. (p. 743)

Material Covered:

____ Example 1: Finding the Volume of a Rectangular Prism

Vocabulary:

volume of a solid, p. 743

____ **Goal 2:** Find the volume of prisms and cylinders in real life. (pp. 744–745)

Material Covered:

____ Example 2: Finding Volumes

____ Example 3: Using Volumes

____ Example 4: Using Volumes in Real Life

Activity 12.4: Minimizing Surface Area (p. 750)

____ **Goal:** Use a spreadsheet to find the minimum surface area of a solid with a given volume.

____ Student Help: Software Help

____ Other (specify) _____

_____

Homework and Additional Learning Support

____ Textbook (specify) <u>pp. 746–749</u> _____

_____

____ *Reteaching with Practice* worksheet (specify exercises)_____

____ *Personal Student Tutor* for Lesson 12.4

NAME _____ DATE _____

Real-Life Application:
When Will I Ever Use This?

For use with pages 743–749

Mailing Packages

Once in a while you may have to send someone a package in the mail. Whether you are sending someone a gift or returning a previously purchased item, you need to pick the appropriate type of packaging. Packaging comes in many different sizes to accommodate your needs. The following are a few packaging tips.

- Select a package that is durable enough to protect its contents.

- Cushion the package contents with packing materials like newspaper and bubble wrap.

- Always use tape that is designed for shipping.

- The delivery and return addresses should be well marked on the package and an index card with the same information should be placed *inside* the package as well.

- Packages that weigh one or more pounds should be taken to the post office for mailing.

In Exercises 1–6, use the following information.

You work part-time at the "Pack-it and Mail-it" that sells containers used for shipping items. Your photography teacher asks you to recommend the best package to ship 35mm film containers. The film containers are 1 inch in diameter and 2 inches in height, and they are to be shipped in groups of twelve. At work you get a list of package sizes. Your store has 4 box sizes that look like they might work. The sizes are given in the form $l \times w \times h$. They are 3 in. × 2 in. × 5 in., 4 in. × 3 in. × 2 in., 5 in. × 6 in. × 5 in., and 3 in. × 2 in. × 3 in. You now have to see which one will fit a dozen film containers with the least amount of wasted space.

1. Find the volume of one film container. Round your result to two decimal places.

2. Find the volume of twelve film containers. Round your result to two decimal places.

3. What is the volume of each package?

4. Which container has a volume that is closest to the volume that you need?

5. Will the package in Exercise 4 work? Why or why not?

6. Which is the next closest size? How can you pack the film containers?

NAME _____ DATE _____

Challenge: Skills and Applications

For use with pages 743–749

1. A silver ingot is in the shape of a right prism whose height is 25 cm and whose base is a trapezoid with height 4.5 cm and one base of length 8 cm. If the mass of the ingot is 8505 g, find the length of the other base of the trapezoid. (The mass of a silver ingot is equal to its volume times 10.5 g/cm³.)

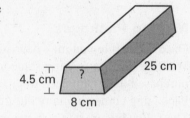

2. Refer to the diagram showing a storage building in the shape of a pentagonal prism. If the building has a volume of 12,000 ft³, find the dimensions of each roof panel.

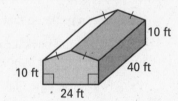

In Exercises 3 and 4, give your answers as decimals rounded to the nearest hundredth.

3. A right circular cylinder has a diameter of 3 in. and a height of 10 in.

   a. Find the volume of the cylinder.

   b. Suppose a right prism whose base is an equilateral triangle is to be placed inside the cylinder. Find the volume of the largest such prism that will fit inside the cylinder.

   c. Now suppose the cylinder is to be placed inside a right prism whose base is an equilateral triangle. Find the volume of the smallest such prism that can be used.

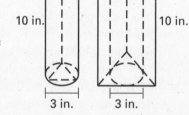

4. A barrel in the shape of a cylinder with radius 2 ft and height 5 ft has been turned on its side, as shown.

   a. If water fills the barrel to a depth of 2 ft, find the volume of the water.

   b. If water fills the barrel to a depth of 1 ft, find the volume of the water. (*Hint:* Use a special right triangle.)

   c. If water fills the barrel to a depth of 3.2 ft, find the volume of the water. (*Hint:* Use trigonometry.)

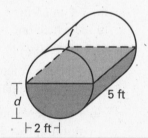

LESSON
12.5

TEACHER'S NAME _____ CLASS _____ ROOM _____ DATE _____

Lesson Plan

2-day lesson (See *Pacing the Chapter,* **TE pages 716C–716D)**   **For use with pages 751–758**

GOALS   1. **Find the volume of pyramids and cones.**
          2. **Find the volume of pyramids and cones in real-life.**

State/Local Objectives _____

_____

✓ Check the items you wish to use for this lesson.

STARTING OPTIONS
____ Homework Check: TE page 746: Answer Transparencies
____ Warm-Up or Daily Homework Quiz: TE pages 752 and 749, CRB page 67, or Transparencies

TEACHING OPTIONS
____ Motivating the Lesson: TE page 753
____ Concept Activity: SE page 751
____ Lesson Opener (Visual Approach): CRB page 68 or Transparencies
____ Examples: Day 1: 1–3, SE pages 752–753; Day 2: 4–5, SE page 754
____ Extra Examples: Day 1: TE page 753 or Transp.; Day 2: TE page 754 or Transp.
____ Closure Question: TE page 754
____ Guided Practice: SE page 755   Day 1: Exs. 1–7; Day 2: Exs. none

APPLY/HOMEWORK
Homework Assignment
____ Basic   Day 1: 8–22; Day 2: 23–37, 40–51; Quiz 2: 1–7
____ Average   Day 1: 8–22; Day 2: 23–37, 40–51; Quiz 2: 1–7
____ Advanced   Day 1: 8–22; Day 2: 23–51; Quiz 2: 1–7

Reteaching the Lesson
____ Practice Masters: CRB pages 69–71 (Level A, Level B, Level C)
____ Reteaching with Practice: CRB pages 72–73 or Practice Workbook with Examples
____ Personal Student Tutor

Extending the Lesson
____ Applications (Interdisciplinary): CRB page 75
____ Challenge: SE page 757; CRB page 76 or Internet

ASSESSMENT OPTIONS
____ Checkpoint Exercises: Day 1: TE page 753 or Transp.; Day 2: TE page 754 or Transp.
____ Daily Homework Quiz (12.5): TE page 758, CRB page 80, or Transparencies
____ Standardized Test Practice: SE page 757; TE page 758; STP Workbook; Transparencies
____ Quiz (12.4–12.5): SE page 758; CRB page 77

Notes _____

_____

_____

LESSON
12.5

TEACHER'S NAME _____ CLASS _____ ROOM _____ DATE _____

Lesson Plan for Block Scheduling

1-day lesson (See *Pacing the Chapter,* TE pages 716C–716D)   **For use with pages 751–758**

GOALS 1. **Find the volume of pyramids and cones.**
2. **Find the volume of pyramids and cones in real-life.**

State/Local Objectives _____

_____

_____

✓ Check the items you wish to use for this lesson.

STARTING OPTIONS
____ Homework Check: TE page 746: Answer Transparencies
____ Warm-Up or Daily Homework Quiz: TE pages 752 and
     749, CRB page 67, or Transparencies

TEACHING OPTIONS
____ Motivating the Lesson: TE page 753
____ Concept Activity: SE page 751
____ Lesson Opener (Visual Approach): CRB page 68 or Transparencies
____ Examples: Day 4: 1–3, SE pages 752–753; Day 5: 4–5, SE page 754
____ Extra Examples: Day 4: TE page 753 or Transp.; Day 5: TE page 754 or Transp.
____ Closure Question: TE page 754
____ Guided Practice: SE page 755   Day 4: Exs. 1–7; Day 5: Exs. none

APPLY/HOMEWORK
Homework Assignment (See also the assignments for Lessons 12.4 and 12.6.)
____ Block Schedule: Day 4: 8–22; Day 5: 25–37, 40–51; Quiz 2: 1–7

Reteaching the Lesson
____ Practice Masters: CRB pages 69–71 (Level A, Level B, Level C)
____ Reteaching with Practice: CRB pages 72–73 or Practice Workbook with Examples
____ Personal Student Tutor

Extending the Lesson
____ Applications (Interdisciplinary): CRB page 75
____ Challenge: SE page 757; CRB page 76 or Internet

ASSESSMENT OPTIONS
____ Checkpoint Exercises: Day 4: TE page 753 or Transp.; Day 5: TE page 754 or Transp.
____ Daily Homework Quiz (12.5): TE page 758, CRB page 80, or Transparencies
____ Standardized Test Practice: SE page 757; TE page 758; STP Workbook; Transparencies
____ Quiz (12.4–12.5): SE page 758; CRB page 77

CHAPTER PACING GUIDE	
Day	**Lesson**
1	Assess Ch. 11; 12.1 (all)
2	12.2 (all); 12.3 (begin)
3	12.3 (end); 12.4 (begin)
4	12.4 (end); **12.5 (begin)**
5	**12.5 (end)**; 12.6 (all)
6	12.7 (all)
7	Review Ch. 12; Assess Ch. 12

Notes _____

_____

_____

Available as a transparency

LESSON
12.5

NAME _____ DATE _____

WARM-UP EXERCISES

For use before Lesson 12.5, pages 751–758

Find the volume of the solid.

1. cube, side lengths of 9

2. cylinder, radius of 3 and height of 8

3. cylinder, diameter of 12 and height of 2

4. rectangular prism, length of 6, width of 11, height of 4

5. triangular prism, base edges of length 4, height of 4

....

DAILY HOMEWORK QUIZ

For use after Lesson 12.4, pages 743–750

1. Find the volume of the right prism.

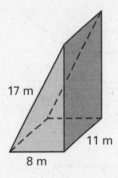

17 m

11 m

8 m

2. Find the volume of a right cylinder that has a diameter of 12 yd and a height of 11 yd. Round the result to one decimal place.

3. A heptagonal prism has a base area of 24 in.^{2} and a height of 5 in. What is its volume?

4. Use Cavalieri's Principle to find the volume of the oblique cylinder. Leave the result in terms of π.

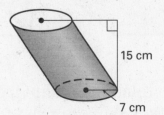

15 cm

7 cm

5. A right rectangular prism has a volume of 385 m^3. Its bases have sides of length 5 m and 7 m. What is its height?

NAME _____ DATE _____

Visual Approach Lesson Opener

For use with pages 752–758

1. Use each of the words *cone*, *cylinder*, *prism*, and *pyramid* twice to complete the statements about the two diagrams below.

   a. A _____ is inscribed in a _____.

   b. A _____ is inscribed in a _____.

   c. _____ is to _____ as _____ is to _____.

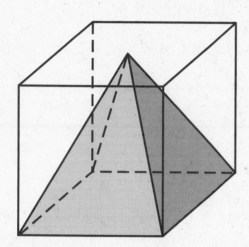 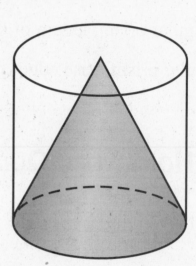

2. Three congruent pyramids fit inside a cube as shown below. The base of each pyramid is one of the faces of the cube. Compare the volume of one pyramid to the volume of the cube.

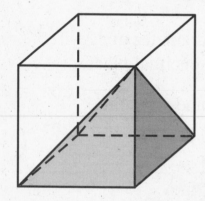 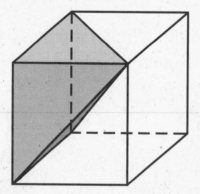 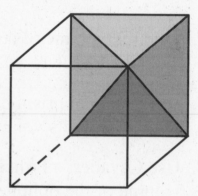

3. Use the relationship in Question 2 to make some conjectures about the volumes of the four solids shown in Exercise 1.

Geometry
Chapter 12 Resource Book

NAME _____ DATE _____

Practice A
For use with pages 752–758

Find the area of the base of the regular pyramid or cone.

1.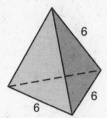
   6
   6
   6
   6

2.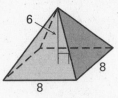
   6
   8
   8

3.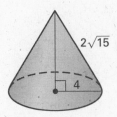
   $2\sqrt{15}$
   4

Find the volume of the pyramid. Each pyramid has a regular polygon for a base.

4.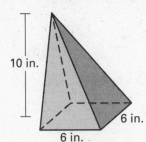
   10 in.
   6 in.
   6 in.

5.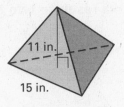
   8 cm
   8 cm
   8 cm

6.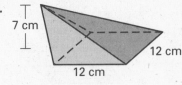
   7 cm
   12 cm
   12 cm

7.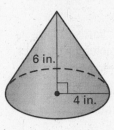
   6 ft
   10 ft

8.
   11 in.
   15 in.

9.
   8 cm
   6 cm

Find the volume of the cone. Round the result to two decimal places.

10.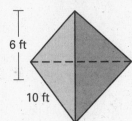
    6 in.
    4 in.

11.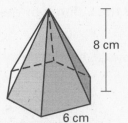
    11 cm
    5 cm

12.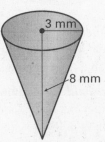
    3 mm
    8 mm

13. **Ice Cream Cone** Find the volume of the ice cream cone shown.

    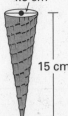
    4.5 cm
    15 cm

14. **Sand** A truck has hauled 48 cubic feet of sand to a building site. If the sand is dumped into a conical shape 4 feet high, what is the diameter?

    4 ft

NAME _____ DATE _____

Practice B

For use with pages 752–758

Find the area of the base of the regular pyramid or cone.

1.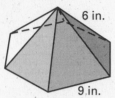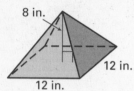
6 in.

9 in.

2.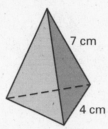
7 cm

4 cm

3.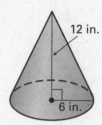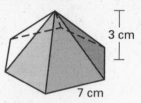
12 in.

6 in.

Find the volume of the pyramid. Each pyramid has a regular polygon for a base.

4.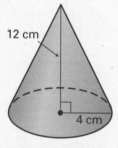
8 in.

12 in.

12 in.

5.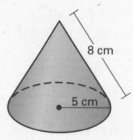
9 in.

5 in.

6.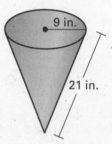
3 cm

7 cm

Find the volume of the cone. Round the result to two decimal places.

7.
12 cm

4 cm

8.
8 cm

5 cm

9.
9 in.

21 in.

10. *Concrete*   To complete a construction job, a contractor needs 145 more cubic yards of concrete. If there remains a conical pile of concrete mix measuring 36 feet in diameter and 12 feet high, is there enough concrete still on the job site to finish the job? Explain your reasoning.

11. *Pyramid*   The limestone blocks from which an ancient pyramid was made weigh about 2 tons per cubic yard. Find the approximate weight of the pyramid having a square base of length 250 yards and a height of 150 yards.

Find the volume of the solid. Each prism is right. Round your result to one decimal place.

12.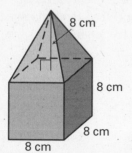
8 cm

8 cm

8 cm

8 cm

13.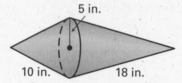
5 in.

10 in.   18 in.

NAME _____ DATE _____

Practice C

For use with pages 752–758

Draw each solid and find its volume.

1. Regular Pyramid
Base: 5 cm by 5 cm
Height: 3 cm

2. Right Cone
Radius: 3 in.
Height: 4 in.

3. Oblique Cone
Diameter: 6 in.
Height: 3 in.

4. Oblique Pyramid
Base: 2 cm by 4 cm
Height: 3 cm

Solve for the variable using the given information. Round to one decimal place.

5. Volume = 1521 cm^3

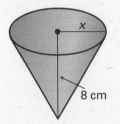

13 cm
13 cm

6. Volume = 170 cm^3

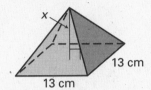

x
8 cm

7. Volume = 81 in.^{3}

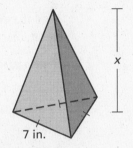

x
7 in.

8. *Jewelry*   A jeweler is casting small gold cones for a special piece of jewelry. The jeweler has 60 grams of gold to use and wishes to make twelve cones with radius 0.5 centimeter and height 1 centimeter. If gold weighs 19.32 grams per cubic centimeter, does the jeweler have enough gold to make the cones? Explain.

9. *Rocket*   A rocket has the dimensions shown at the right. If 60% of the space in the rocket is needed for fuel, what is the volume, to the nearest whole unit, of the portion of the rocket that is available for nonfuel items?

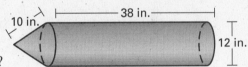

10 in.   38 in.   12 in.

10. *Gazebo*   A gazebo has a pentagonal base with an area of 80 square meters. The total height to the peak is 8 meters. The height of the pyramidal roof is 2 meters. Find the gazebo's total volume.

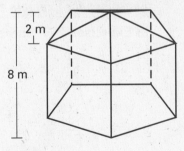

2 m
8 m

Find the volume of the solid. Each prism is right. Round your result to one decimal place.

11.

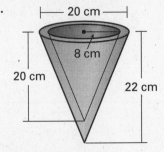

20 cm
8 cm
20 cm
22 cm

12.

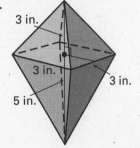

3 in.
3 in.
3 in.
5 in.

NAME _____ DATE _____

Reteaching with Practice

For use with pages 752–758

GOAL Find the volume of pyramids and cones

VOCABULARY

Theorem 12.9 Volume of a Pyramid   The volume V of a pyramid is $V = \frac{1}{3}Bh$, where B is the area of the base and h is the height.

Theorem 12.10 Volume of a Cone   The volume V of a cone is $V = \frac{1}{3}Bh = \frac{1}{3}\pi r^2 h$, where B is the area of the base, h is the height, and r is the radius of the base.

EXAMPLE 1 *Finding the Volume of a Pyramid*

Find the volume of the pyramid with the square base shown to the right.

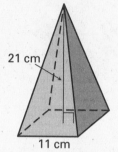

SOLUTION

The area B of the base of the pyramid is the area of the square. Using the formula for the area of a square, s^2, $B = 11^2$, or 121 square centimeters. Using $h = 21$, you can find the volume.

$$V = \frac{1}{3}Bh \qquad \text{Formula for volume of pyramid}$$

$$= \frac{1}{3}(121)(21) \qquad \text{Substitute.}$$

$$= 847 \qquad \text{Simplify.}$$

So, the volume of the pyramid is 847 cubic centimeters.

Exercises for Example 1

In Exercises 1–3, find the volume of the pyramid.

1.
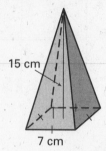

2.
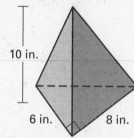

3.
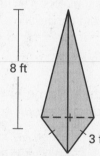

Geometry
Chapter 12 Resource Book

NAME _____ DATE _____

Reteaching with Practice

For use with pages 752–758

EXAMPLE 2 ***Finding the Volume of a Cone***

Find the volume of the cone.

SOLUTION

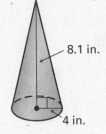
8.1 in.

4 in.

$$V = \frac{1}{3}Bh = \frac{1}{3}(\pi r^2)h \qquad \text{Formula for volume of cone}$$

$$= \frac{1}{3}(\pi \cdot 4^2)(8.1) \qquad \text{Substitute.}$$

$$= 43.2\pi \qquad \text{Simplify.}$$

So, the volume of the cone is 43.2π in.^{3}, or about 135.7 in.^{3}.

Exercises for Example 2

Find the volume of the cone.

4.

4 in.
15 in.

5. 9.2 m
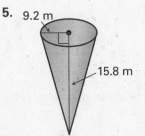
15.8 m

6.
26 m
19.5 m

EXAMPLE 3 ***Using the Volume of a Cone***

Use the given measurements to solve for x.

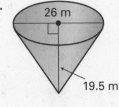
$V = 105$ cm^3
x
5 cm

SOLUTION

$$V = \frac{1}{3}\pi r^2 h \qquad \text{Formula for volume of cone}$$

$$105 = \frac{1}{3}\pi \cdot 5^2 \cdot x \qquad \text{Substitute.}$$

$$4 \approx x \qquad \text{Simplify and solve for } x.$$

The height of the cone is about 4 centimeters.

Exercises for Example 3

In Exercises 7–9, find the value of x.

7.
$V = 182$ in.^{3}

x
6 in.

8. $V = 215$ m^3
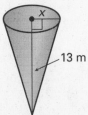
x
13 m

9. $V = 56.5$ m^3
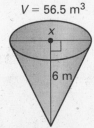
x
6 m

NAME _____ DATE _____

Quick Catch-Up for Absent Students

For use with pages 751–758

The items checked below were covered in class on (date missed) _____

Activity 12.5: Investigating Volume (p. 751)

_____ **Goal:** Determine how the volume of a pyramid is related to the volume of a prism with the same base and height.

Lesson 12.5: Volume of Pyramids and Cones

_____ **Goal 1:** Find the volume of pyramids and cones. (pp. 752–753)

Material Covered:

_____ Example 1: Finding the Volume of a Pyramid

_____ Student Help: Study Tip

_____ Example 2: Finding the Volume of a Cone

_____ Student Help: Study Tip

_____ Example 3: Using the Volume of a Cone

_____ **Goal 2:** Find the volume of pyramids and cones in real life. (p. 754)

Material Covered:

_____ Example 4: Finding the Volume of a Solid

_____ Example 5: Using the Volume of a Cone

_____ Other (specify) _____

_____

Homework and Additional Learning Support

_____ Textbook (specify) _pp. 755–758_____

_____

_____ *Reteaching with Practice* worksheet (specify exercises)_____

_____ *Personal Student Tutor* for Lesson 12.5

LESSON

12.5

For use with pages 752–758

Lesson 12.5

NAME _____ DATE _____

Interdisciplinary Application

The Great Pyramid

HISTORY   The Great Pyramid in Giza is one of the Seven Wonders of
the Ancient World. It is 756 feet along its square base and is 450 feet
tall. Built about 4,500 years ago, it is the only one of the ancient
wonders that is still standing. It has a greater volume than the Empire
State Building and until the 1800s it was the tallest building in the
world. There are about 2,300,000 stones in the pyramid, that were
dragged to the top by teams of workers up ramps of clay built up
around the pyramids. Historians have uncovered evidence that the
workers were not slaves, as once thought, but paid workers that worked
in the agricultural flood season when the banks of the Nile would
overflow and no work could be done in the fields.

In Exercises 1–4, use the information above.

1. What is the volume of the Great Pyramid?

2. If each block were 3 ft × 3 ft × 4 ft, what would the volume of each
   block be?

3. How many cubic feet of stone are in the Great Pyramid?

4. How much of the volume is left for the passages and chambers
   inside?

NAME _____ DATE _____

Challenge: Skills and Applications

For use with pages 752–758

1. The figure shown is a regular octahedron. Each face is an equilateral triangle with a side length of $3\sqrt{2}$ in. Find the volume of the regular octahedron.

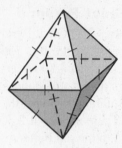

2. A *cuboctahedron* has 6 square faces and 8 equilateral triangle faces. It can be made by slicing off the corners of a cube, as shown. If each edge of a cuboctahedron has length $3\sqrt{2}$ cm, find the volume of the cuboctahedron. (*Hint:* Find the volume of the original cube, and subtract the volume of the corners that are removed.)

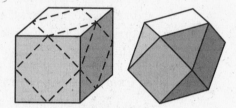

3. The *frustum* of a pyramid or cone is obtained by slicing off the top portion of the pyramid or cone, as shown, where the cut is parallel to the base of the pyramid or cone. Let B be the area of the base ($PQRS$), and let C be the area of the cut surface ($TUVW$). Let h be the height of the frustum, and let k be the height of the small pyramid (or cone).

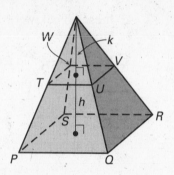

a. Show that the volume V of the frustum is given by
$$V = \frac{1}{3}Bh + \frac{1}{3}(B - C)k.$$

b. The cut surface is similar to the base of the original pyramid or cone, and the side lengths are in the ratio $k : h + k$. Use areas of similar figures to show that $\dfrac{h}{k} + 1 = \sqrt{\dfrac{B}{C}}$.

c. Show that $\dfrac{k}{h} = \dfrac{C + \sqrt{BC}}{B - C}$.

d. Show that $V = \dfrac{h}{3}\left(B + C + \sqrt{BC}\right)$.

4. In the diagram for Exercise 3, suppose the area of $PQRS$ is 50 ft² and the area of $TUVW$ is 8 ft². Let $h = 3$ ft.

a. Use your result from Exercise 3 to find the volume of the frustum.

b. Find k. Then subtract the volume of the small pyramid from the volume of the large pyramid to verify your answer to part (a).

NAME _____ DATE _____

Quiz 2

For use after Lessons 12.4 and 12.5

In Exercises 1–6, find the volume of the solid.
(Lessons 12.4 and 12.5)

1.

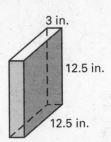

3 in.

12.5 in.

12.5 in.

2.

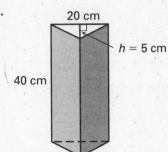

20 cm

$h = 5$ cm

40 cm

3.

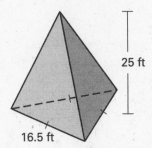

15 m   30 m

4.

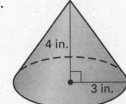

4 in.

3 in.

5.

25 ft

16.5 ft

6.

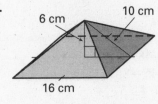

6 cm   10 cm

16 cm

7. In a certain cone, the radius of the base has the same length as the altitude of the cone. The length of the slant height of the cone is 6 centimeters. Find the volume of the cone.

TEACHER'S NAME _____ CLASS _____ ROOM _____ DATE _____

Lesson Plan

1-day lesson (See *Pacing the Chapter,* TE pages 716C–716D)   For use with pages 759–765

GOALS 1. **Find the surface area of a sphere.**
2. **Find the volume of a sphere in real life.**

State/Local Objectives _____

_____

✓ **Check the items you wish to use for this lesson.**

STARTING OPTIONS
_____ Homework Check: TE page 755: Answer Transparencies
_____ Warm-Up or Daily Homework Quiz: TE pages 759 and 758, CRB page 80, or Transparencies

TEACHING OPTIONS
_____ Motivating the Lesson: TE page 760
_____ Lesson Opener (Technology): CRB page 81 or Transparencies
_____ Technology Activity with Keystrokes: CRB pages 82–85
_____ Examples 1–4: SE pages 759–761
_____ Extra Examples: TE pages 760–761 or Transparencies; Internet
_____ Closure Question: TE page 761
_____ Guided Practice Exercises: SE page 762

APPLY/HOMEWORK
Homework Assignment
_____ Basic   10–17, 20–29, 35, 36, 45–48, 50–57
_____ Average   10–18, 20–29, 35–40, 45–48, 50–57
_____ Advanced   10–18, 20–29, 35–40, 44–57

Reteaching the Lesson
_____ Practice Masters: CRB pages 86–88 (Level A, Level B, Level C)
_____ Reteaching with Practice: CRB pages 89–90 or Practice Workbook with Examples
_____ Personal Student Tutor

Extending the Lesson
_____ Applications (Real-Life): CRB page 92
_____ Challenge: SE page 765; CRB page 93 or Internet

ASSESSMENT OPTIONS
_____ Checkpoint Exercises: TE pages 760–761 or Transparencies
_____ Daily Homework Quiz (12.6): TE page 765, CRB page 96, or Transparencies
_____ Standardized Test Practice: SE page 765; TE page 765; STP Workbook; Transparencies

Notes _____

_____

_____

TEACHER'S NAME _____ CLASS _____ ROOM _____ DATE _____

Lesson Plan for Block Scheduling

Half-day lesson (See *Pacing the Chapter,* TE pages 716C–716D)   **For use with pages 759–765**

GOALS   1. **Find the surface area of a sphere.**
2. **Find the volume of a sphere in real life.**

State/Local Objectives _____

_____

_____

✓ Check the items you wish to use for this lesson.

STARTING OPTIONS

____ Homework Check: TE page 755: Answer Transparencies
____ Warm-Up or Daily Homework Quiz: TE pages 759 and
         758, CRB page 80, or Transparencies

TEACHING OPTIONS

____ Motivating the Lesson: TE page 760
____ Lesson Opener (Technology): CRB page 81 or Transparencies
____ Technology Activity with Keystrokes: CRB pages 82–85
____ Examples 1–4: SE pages 759–761
____ Extra Examples: TE pages 760–761 or Transparencies; Internet
____ Closure Question: TE page 761
____ Guided Practice Exercises: SE page 762

APPLY/HOMEWORK
Homework Assignment (See also the assignment for Lesson 12.5.)
____ Block Schedule: 10–18, 20–29, 35–40, 45–48, 50–57

Reteaching the Lesson
____ Practice Masters: CRB pages 86–88 (Level A, Level B, Level C)
____ Reteaching with Practice: CRB pages 89–90 or Practice Workbook with Examples
____ Personal Student Tutor

Extending the Lesson
____ Applications (Real-Life): CRB page 92
____ Challenge: SE page 765; CRB page 93 or Internet

ASSESSMENT OPTIONS

____ Checkpoint Exercises: TE pages 760–761 or Transparencies
____ Daily Homework Quiz (12.6): TE page 765, CRB page 96, or Transparencies
____ Standardized Test Practice: SE page 765; TE page 765; STP Workbook; Transparencies

CHAPTER PACING GUIDE	
Day	**Lesson**
1	Assess Ch. 11; 12.1 (all)
2	12.2 (all); 12.3 (begin)
3	12.3 (end); 12.4 (begin)
4	12.4 (end); 12.5 (begin)
5	12.5 (end); **12.6 (all)**
6	12.7 (all)
7	Review Ch. 12; Assess Ch. 12

Lesson 12.6

Notes _____

_____

_____

NAME _____ DATE _____

WARM-UP EXERCISES

For use before Lesson 12.6, pages 759–765

Find the volume of the figure.

1. right cylinder, radius of 10 and height of 4

2. right cone, base diameter of 18, height of 8

Find the surface area of the figure.

3. right cylinder, radius of 9 and height of 11

4. right cone, radius of 5, slant height of 6

....

DAILY HOMEWORK QUIZ

For use after Lesson 12.5, pages 751–758

1. Find the volume of the pyramid
with the regular base.

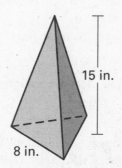

15 in.

8 in.

2. A right cone has a radius of 6 ft and a slant height of 9 ft.
What is its volume? Round the result to two decimal places.

3. A cone has a diameter of 10 m and a volume of 150π m³.
What is its height?

4. Find the volume of the solid.

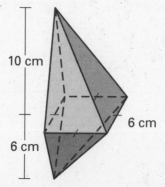

10 cm

6 cm

6 cm

6 cm

Technology Lesson Opener

For use with pages 759–765

Use a calculator or a spreadsheet to compare the surface area of each planet to the surface of Earth.

1. The surface area S of a sphere with radius r is $S = 4\pi r^2$. Use this formula to find the surface area of each planet rounded to the nearest million square miles. Calculate the indicated ratio to complete the table to the nearest hundredth.

Planet	Diameter at equator (mi)	Surface area (mi ^{2})	$\dfrac{\text{Surface area of planet}}{\text{Surface area of Earth}}$
Mercury	3026		
Venus	7504		
Earth	7909		1
Mars	4212		
Jupiter	88,650		
Saturn	74,732		
Uranus	31,693		
Neptune	30,710		
Pluto	1410		

2. For each planet besides Earth, use the ratio you calculated to write a sentence comparing the surface area of the planet to the surface area of Earth. Use language and forms of numbers, such as percents, that make the comparison easy to understand.

NAME _____ DATE _____

Technology Activity

For use with pages 759–765

GOAL To use a spreadsheet to find the relationships, if any, that exist between the radius of a sphere and the sphere's surface area and volume.

Activity

❶ Make a table with six columns with the headers Radius, Surface Area, Volume, Ratio of the Radii, Ratio of the Surface Areas, and Ratio of the Volumes.

❷ In cell A2 enter a value of 1.

❸ In cell A3 enter the formula =A2+1 and use the Fill Down feature to create values up to 10.

❹ In cell B2 enter the formula =4*PI()*A2^2 as the formula for the surface area. Your spreadsheet might use a different formula for π.

❺ In cell C2 use the formula =(4/3)*PI()*A2^3 as the formula for the volume.

❻ Use the Fill Down feature to determine values in columns B and C.

❼ In cell D3 use the formula =A3/$A$2. The results in cells D3–D11 compare the radius of that row's sphere with a radius of 1 unit.

❽ In cell E3 use the formula =B3/$B$2. Use the Fill Down feature in column E. The results in cells E3–E11 compare the surface area of that row's sphere with the surface area of a sphere with a radius of 1 unit.

❾ In cell F3 use the formula =C3/$C$2. Use the Fill Down feature in column F. The results in cells F3–F11 compare the volume of that row's sphere with the volume of a sphere with a radius of 1 unit.

Exercises

1. When the radius of the sphere was doubled from 1 unit to 2 units, how was the resulting surface area affected? How was the resulting volume affected?

2. Look at the other rows in the table. For example, when the radius of the sphere was multiplied by 5, how was the resulting surface area affected? How was the resulting volume affected?

3. Look for patterns in the table to complete the following conjecture. If the radius of a sphere is multiplied by n, then the resulting surface area is multiplied by __?__ and the resulting volume is multiplied by __?__ .

LESSON
12.6
CONTINUED

NAME _____ DATE _____

Technology Activity

For use with pages 759–765

EXCEL

1. Select cell A1.

   Radius `TAB` Surface Area `TAB` Volume `TAB` Ratio of the Radii `TAB`

   Ratio of the Surface Areas `TAB` Ratio of the Volumes `ENTER`

2. Select cell A2.

   1 `ENTER`

3. Select cell A3.

   =A2+1 `ENTER`

   Select cells A3–A11. From the **Edit** menu, choose **Fill Down**.

4. Select cell B2.

   =4*PI()*A2^2 `TAB`

5. Select cell C2.

   =(4/3)*PI()*A2^3 `ENTER`

6. Select cells B2–B11 and C2–C11. From the **Edit** menu, choose **Fill Down**.

LESSON

12.6

CONTINUED

NAME _____ DATE _____

Technology Activity

For use with pages 759–765

7. Select cell D3.

=A3/$A$2 [ENTER]

Select cells D3–D11. From the **Edit** menu, choose **Fill Down**.

8. Select cell E3.

=B3/$B$2 [ENTER]

Select cells E3–E11. From the **Edit** menu, choose **Fill Down**.

The results in cells E3–E11 compare the surface area of the sphere in that row to the surface area of a sphere with a radius of 1 unit.

9. Select cell F3.

=C3/$C$2 [ENTER]

Select cells F3–F11. From the **Edit** menu, choose **Fill Down**.

The results in cells F3–F11 compare the volume of the sphere in that row to the volume of a sphere with a radius of 1 unit.

Lesson 12.6

NAME _____ DATE _____

Technology Activity Keystrokes

For use with page 763

Keystrokes for Exercises 30–32
TI-92

1. APPS 2

   Cursor to y 1.

   ENTER (4 ÷ 3) 2nd [π] x ^ 3 ENTER

   ENTER

   4 2nd [π] x ^ 2 ENTER ENTER y 1 (x) ÷

   (

   y 2 (x))

2. ◆ [TblSet] 1 ▼ 1 ▼ ▶ 1 ▶ ▼ ▶ 1 ENTER

3. APPS 5

4. APPS 2

   Turn off y1 and y2 (F4). Then graph y3.

   F2 6

Practice A

For use with pages 759–765

Use the diagram at the right.

1. Name a chord of the sphere.

2. Name a segment that is a radius of the sphere.

3. Name a segment that is a diameter of the sphere.

4. Find the circumference of the great circle that contains *A* and *B*.

5. Find the surface area of the sphere.

6. Find the volume of the sphere.

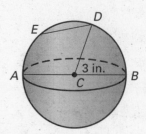

Find the surface area of the sphere. Round your result to two decimal places.

7.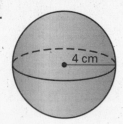

8.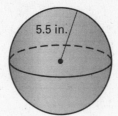

9.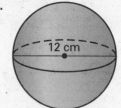

In Exercises 10–13, use the diagram at the right. The center of the sphere is *C* and its circumference is 9π inches.

10. What is half of the sphere called?

11. Find the radius of the sphere.

12. Find the diameter of the sphere.

13. Find the surface area of half the sphere.

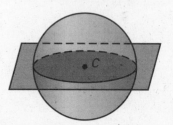

Find the volume of the sphere. Round the result to two decimal places.

14.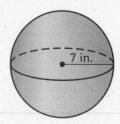

15.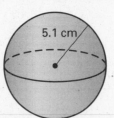

16.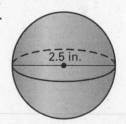

17. *Earth and Mercury* The mean radius of Earth is approximately 3963 miles. The mean radius of Mercury is 1509 miles. How does the surface area of Mercury compare to the surface area of Earth?

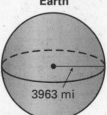

NAME _____ DATE _____

Practice B

For use with pages 759–765

Find the surface area and the volume of the sphere.

1.

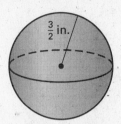

$\frac{3}{2}$ in.

2.

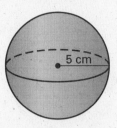

5 cm

3.

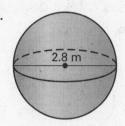

2.8 m

Complete the table below. Leave your answers in terms of π.

	Radius of sphere	Circumference of great circle	Surface area of sphere	Volume of sphere
4.	10 mm	_____	_____	_____
5.	_____	36π in.	_____	_____
6.	_____	_____	2304π cm^2	_____
7.	_____	_____	_____	$\frac{500}{3}\pi$ yd^3

Find the surface area of the solid and the volume of the solid. The cylinder and cone are right. Round your results to two decimal places.

8.

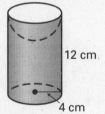

12 cm

4 cm

9.

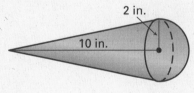

2 in.

10 in.

Tennis Ball   **In Exercises 10 and 11, consider a tennis ball with a radius of 3.2 centimeters.**

10. Find the surface area and volume of the tennis ball.

11. Tennis balls are often sold in a can of three. Assuming that the balls are packed tightly so that they touch the lateral side and the bases, determine the amount of volume in the can that is not taken up by the tennis balls.

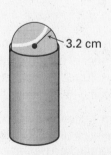

3.2 cm

Earth   **In Exercises 12 and 13, Earth has an equatorial radius of approximately 3963 miles.**

12. Seventy percent of Earth's surface is covered with water. Find the approximate surface area of Earth that is water.

13. Find the volume of Earth.

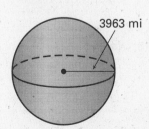

3963 mi

NAME _____ DATE _____

Practice C
For use with pages 759–765

Find the surface area and the volume of the sphere. Round your answers to two decimal places.

1.

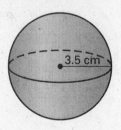

3.5 cm

2.

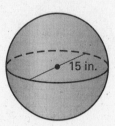

15 in.

3.

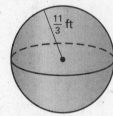

$\frac{11}{3}$ ft

Complete the table below. Leave your answers in terms of π.

	Radius of sphere	Circumference of great circle	Surface area of sphere	Volume of sphere
4.	15 cm	_____	_____	_____
5.	_____	26π ft	_____	_____
6.	_____	_____	4096π in.^{2}	_____
7.	_____	_____	_____	$\dfrac{1372}{3}\pi$ yd^3

Find the area of the intersection of the sphere and plane. Leave your answers in terms of π.

8.

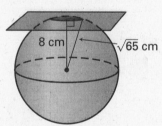

8 cm $\sqrt{65}$ cm

9.

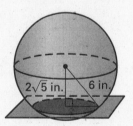

$2\sqrt{5}$ in.   6 in.

10.

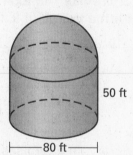

2 m

$2\sqrt{3}$ m

11. *Storage Tank* A grain storage tank shown at the right is in the shape of a cylinder covered by a half sphere. If the height of the cylinder is 50 feet and it is 80 feet in diameter, find the total surface area (including the base) and volume of the tank.

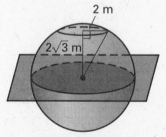

50 ft

80 ft

12. *Box* How much extra space will you need to fill the inside of a box with dimensions 44 cm × 44 cm × 44 cm after placing a bowling ball inside with a radius of 21.8 cm?

13. *Rubber Ball* A rubber shell filled with air forms a rubber ball. The shell's outer diameter is 65 millimeters, and its inner diameter is 56 millimeters. Find the volume of rubber used to make the ball. Round your answer to the nearest cubic centimeter.

NAME _____ DATE _____

Reteaching with Practice

For use with pages 759–765

GOAL **Find the surface area of a sphere and find the volume of a sphere**

VOCABULARY

A **sphere** is the locus of points in space that are a given distance from a point called the **center of the sphere**.

A **radius of a sphere** is a segment from the center to a point on the sphere.

A **chord of a sphere** is a segment whose endpoints are on the sphere.

A **diameter of a sphere** is a chord that contains the center.

If a plane that intersects a sphere contains the center of the sphere, the intersection is a **great circle** of the sphere.

A great circle of a sphere separates the sphere into two congruent halves called **hemispheres**.

Theorem 12.11 Surface Area of a Sphere The surface area S of a sphere with radius r is $S = 4\pi r^2$.

Theorem 12.12 Volume of a Sphere The volume V of a sphere with radius r is $V = \frac{4}{3}\pi r^3$.

EXAMPLE 1 *Finding the Surface Area of a Sphere*

Find the surface area of the sphere.

SOLUTION

$$S = 4\pi r^2 \qquad \text{Formula for surface area of sphere}$$

$$= 4\pi(10)^2 \qquad \text{Substitute.}$$

$$= 400\pi \qquad \text{Simplify.}$$

So, the surface area of the sphere is 400π square feet, or about 1256.6 square feet.

Exercises for Example 1

Find the surface area of the sphere.

1.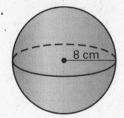

2.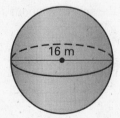

3.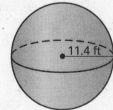

NAME _____ DATE _____

Reteaching with Practice

For use with pages 759–765

EXAMPLE 2 *Using a Great Circle*

The circumference of a great circle of a sphere
is 25 inches. Find the surface area of the sphere.

SOLUTION

Begin by finding the radius of the sphere.

$C = 2\pi r$ Formula for circumference of a circle

$25 = 2\pi r$ Substitute.

$4 \approx r$ Divide each side by 2π.

Using a radius of 4 cm, the surface area is $S = 4\pi r^2 = 4\pi(4)^2 = 64\pi$ in.^{2}

So, the surface area of the sphere is 64π in.^{2}, or about 201.1 in.^{2}

Exercises for Example 2

Find the surface area of the sphere.

4.
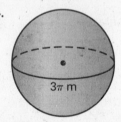
3π m

5.
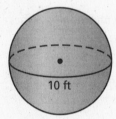
10 ft

6.
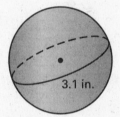
3.1 in.

EXAMPLE 3 *Finding the Volume of a Sphere*

Find the volume of the sphere.

SOLUTION

$V = \dfrac{4}{3}\pi r^3$ Formula for volume of sphere

$= \dfrac{4}{3}\pi(3.5)^3$ Substitute.

≈ 179.6 Simplify.

3.5 ft

So, the volume of the sphere is about 179.6 cubic feet.

Exercises for Example 3

Find the volume of the sphere.

7.
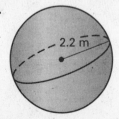
2.2 m

8.
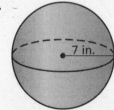
7 in.

9.
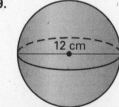
12 cm

NAME _____ DATE _____

Quick Catch-Up for Absent Students

For use with pages 759–765

The items checked below were covered in class on (date missed) _____

Lesson 12.6: Surface Area and Volume of Spheres

_____ **Goal 1:** Find the surface area of a sphere. (pp. 759–760)

Material Covered:

_____ Example 1: Finding the Surface Area of a Sphere

_____ Example 2: Using a Great Circle

_____ Example 3: Finding the Surface Area of a Sphere

Vocabulary:

sphere, p. 759   center of a sphere, p. 759

radius of a sphere, p. 759   chord of a sphere, p. 759

diameter, p. 759   great circle, p. 760

hemisphere, p. 760

_____ **Goal 2:** Find the volume of a sphere in real life. (p. 761)

Material Covered:

_____ Student Help: Study Tip

_____ Example 4: Finding the Volume of a Sphere

_____ Other (specify) _____

_____

Homework and Additional Learning Support

_____ Textbook (specify) _pp. 762–765_____

_____

_____ Internet: Extra Examples at www.mcdougallittell.com

_____ *Reteaching with Practice* worksheet (specify exercises)_____

_____ *Personal Student Tutor* for Lesson 12.6

NAME _____ DATE _____

Real-Life Application:
When Will I Ever Use This?

For use with pages 759–765

The Peachoid

Off Interstate 85 in Gaffney, South Carolina, there is a large spherical water tank shaped and painted to look like a peach. It was built in 1980–81. Although peaches are often associated with Georgia, South Carolina produces more peaches, and Gaffney is home of the South Carolina Peach Festival. The peach-shaped water tank called the "Peachoid" holds one million gallons of water.

In Exercises 1–3, use the information above.

1. Find the radius, diameter, and circumference of the Peachoid in feet. One gallon is equal to 0.1337 cubic feet. Round your results to two decimal places.

2. Find the surface area.

3. The company that made the steel framework for the Peachoid had an agreement with the city of Gaffney that they would not build another peachoid for fifteen years. Now, there is a peachoid tank off Interstate 65 in Clanton, Alabama. It only holds a half-million gallons. How does the radius of the Clanton tank compare to the radius of the tank in Gaffney?

Challenge: Skills and Applications
For use with pages 759–765

1. Suppose that three spherical balls of wax of radius r are packed into a cylinder, also of radius r, as shown. If the wax is melted down, to what depth will the can be filled? Give your answer in terms of r.

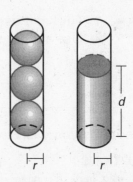

2. If a hole of radius r is drilled through the center of a sphere of radius R, we refer to the remaining portion of the sphere as a *bead* with *inner radius* r and *outer radius* R. The *height* of the bead is h, as shown. It can be shown (using calculus) that the volume of a bead depends only on its height. Since the volume is the same whatever the value of r, you can find the answer to parts (a) and (b) by thinking of the special case where $r = 0$ and the bead is actually a sphere.

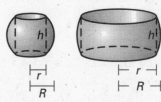

   a. What is the volume of a bead with height 12 cm?

   b. Give a formula for the volume V of a bead in terms of the height h.

3. Suppose a spherical bowl of radius R is filled with liquid to a depth d, where $d < R$. Complete the following steps to find a formula for the volume of the liquid. Let r be the radius of the circle that forms the top surface of the liquid.

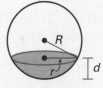

   a. Explain why the height of the bead with outer radius R and inner radius r is $2R - 2d$. Then use your formula from Exercise 2(b) to find the volume of the bead in terms of R and d.

   b. Refer to the cylinder shown at the right. First find the length of the unknown leg of the triangle in terms of R and d. Next use the Pythagorean Theorem to find r in terms of R and d. Then find the volume of the cylinder whose radius is r and whose height is $2R - 2d$.

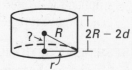

   c. If the volumes you found in parts (a) and (b) are subtracted from $\frac{4}{3}\pi R^3$, the result is *twice* the volume of the liquid.

   Use this fact to show that the volume of the liquid is $V = \frac{\pi d^2}{3}(3R - d)$.

   (*Note:* Although we assumed that $d < R$ to derive this formula, the formula is actually valid for $0 \le d \le 2R$.)

4. Use the formula from Exercise 3(c) to find the volume of the liquid for each value of d. Make a sketch of the bowl with the given amount of liquid and tell whether the volume makes sense.

   a. $d = 0$ b. $d = R$ c. $d = 2R$

5. If $d = \frac{R}{2}$, the volume of the liquid is what fraction of the volume of the sphere?

   Does your result seem reasonable? Explain.

TEACHER'S NAME _____ CLASS _____ ROOM _____ DATE _____

Lesson Plan

2-day lesson (See *Pacing the Chapter,* TE pages 716C–716D)   For use with pages 766–772

GOALS   1. **Find and use the scale factor of similar solids.**
         2. **Use similar solids to solve real-life problems.**

State/Local Objectives _____

_____

✓ **Check the items you wish to use for this lesson.**

STARTING OPTIONS

____ Homework Check: TE page 762: Answer Transparencies
____ Warm-Up or Daily Homework Quiz: TE pages 766 and 765, CRB page 96, or Transparencies

TEACHING OPTIONS

____ Lesson Opener (Activity): CRB page 97 or Transparencies
____ Examples: Day 1: 1–4, SE pages 766–768; Day 2: 5, SE page 768
____ Extra Examples: Day 1: TE pages 767–768 or Transp.; Day 2: TE page 768 or Transp.
____ Closure Question: TE page 768
____ Guided Practice: SE page 769   Day 1: Exs. 1–8; Day 2: Exs. none

APPLY/HOMEWORK

Homework Assignment

____ Basic   Day 1: 10–24 even, 25–29, 39–44; Day 2: 9–23 odd, 30–36, 45–49; Quiz 3: 1–8
____ Average   Day 1: 10–24 even, 25–29, 39–44; Day 2: 9–23 odd, 30–36, 45–49; Quiz 3: 1–8
____ Advanced   Day 1: 10–24 even, 25–29, 39–44; Day 2: 9–23 odd, 30–38, 45–49; Quiz 3: 1–8

Reteaching the Lesson

____ Practice Masters: CRB pages 98–100 (Level A, Level B, Level C)
____ Reteaching with Practice: CRB pages 101–102 or Practice Workbook with Examples
____ Personal Student Tutor

Extending the Lesson

____ Cooperative Learning Activity: CRB page 104
____ Applications (Interdisciplinary): CRB page 105
____ Challenge: SE page 771; CRB page 106 or Internet

ASSESSMENT OPTIONS

____ Checkpoint Exercises: Day 1: TE pages 767–768 or Transp.; Day 2: TE page 768 or Transp.
____ Daily Homework Quiz (12.7): TE page 772, or Transparencies
____ Standardized Test Practice: SE page 771; TE page 772; STP Workbook; Transparencies
____ Quiz (12.6–12.7): SE page 772

Notes _____

_____

_____

TEACHER'S NAME _____ CLASS _____ ROOM _____ DATE _____

Lesson Plan for Block Scheduling
1-day lesson (See *Pacing the Chapter,* TE pages 716C–716D)   For use with pages 766–772

GOALS   1. Find and use the scale factor of similar solids.
2. Use similar solids to solve real-life problems.

State/Local Objectives _____

_____

_____

✓ **Check the items you wish to use for this lesson.**

STARTING OPTIONS
____ Homework Check: TE page 762: Answer Transparencies
____ Warm-Up or Daily Homework Quiz: TE pages 766 and
       765, CRB page 96, or Transparencies

TEACHING OPTIONS
____ Lesson Opener (Activity): CRB page 97 or Transparencies
____ Examples 1–5: SE pages 766–768
____ Extra Examples: TE pages 767–768 or Transparencies
____ Closure Question: TE page 768
____ Guided Practice Exercises: SE page 769

APPLY/HOMEWORK
Homework Assignment
____ Block Schedule: 9–36, 39–49; Quiz 3: 1–8

Reteaching the Lesson
____ Practice Masters: CRB pages 98–100 (Level A, Level B, Level C)
____ Reteaching with Practice: CRB pages 101–102 or Practice Workbook with Examples
____ Personal Student Tutor

Extending the Lesson
____ Cooperative Learning Activity: CRB page 104
____ Applications (Interdisciplinary): CRB page 105
____ Challenge: SE page 771; CRB page 106 or Internet

ASSESSMENT OPTIONS
____ Checkpoint Exercises: TE pages 767–768 or Transparencies
____ Daily Homework Quiz (12.7): TE page 772, or Transparencies
____ Standardized Test Practice: SE page 771; TE page 772; STP Workbook; Transparencies
____ Quiz (12.6–12.7): SE page 772

CHAPTER PACING GUIDE	
Day	Lesson
1	Assess Ch. 11; 12.1 (all)
2	12.2 (all); 12.3 (begin)
3	12.3 (end); 12.4 (begin)
4	12.4 (end); 12.5 (begin)
5	12.5 (end); 12.6 (all)
6	**12.7 (all)**
7	Review Ch. 12; Assess Ch. 12

Lesson 12.7

Notes _____

_____

_____

NAME _____ DATE _____

WARM-UP EXERCISES

For use before Lesson 12.7, pages 766–772

1. Find the surface area and volume of a square prism with a base edge of 1 and a height of 2.

2. Find the surface area and volume of a square prism with a base edge of 3 and a height of 6.

3. Find the surface area and volume in terms of π of a right cone with a radius of 3 and a height of 4.

···

DAILY HOMEWORK QUIZ

For use after Lesson 12.6, pages 759–765

1. Find the surface area of a sphere with a radius of 5 m. Round the result to two decimal places.

2. Find the volume of a sphere with a diameter of 9.5 ft. Round the result to two decimal places.

3. Find the surface area and volume of a sphere with a radius of 3 cm. Leave the results in terms of π.

4. Solve for the variable. Then find the area of the intersection of the sphere and the plane.

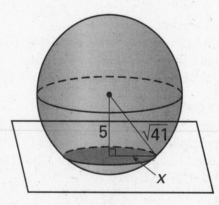

NAME _____ DATE _____

Activity Lesson Opener

For use with pages 766–772

SET UP: Work in a group.
YOU WILL NEED: • box of sugar cubes

1. Each group uses sugar cubes to build successively bigger cubes.
    Start with one sugar cube, and let its side length be 1 unit. Then
    build cubes with side lengths of 2 units, 3 units, 4 units, and
    5 units. As soon as you build a cube, find its surface area and
    volume to complete the table. How is the volume related to the
    number of sugar cubes used?

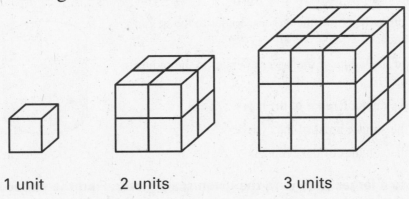

| 1 unit | 2 units | 3 units |

Side length (units)	Surface area (square units)	Volume (cubic units)
1		
2		
3		
4		
5		

2. Look for a pattern in the numbers in the Surface area column.
    Write a formula for the surface area of a cube with side length n.

3. Look for a pattern in the numbers in the Volume column.
    Write a formula for the volume of a cube with side length n.

4. As the side length of a cube increases by a factor of n, how does
    the surface area increase? How does the volume increase?

NAME _____ DATE _____

Practice A
For use with pages 766–772

Decide whether the solids are similar. If so, determine the scale factor.

1.
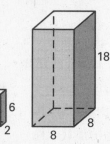

2.
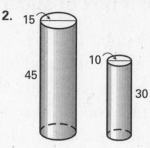

In Exercises 3–7, use the diagram at the right.

3. What is the ratio of the height of the larger cylinder to the height of the smaller cylinder?

4. What is the ratio of the radius of the larger cylinder to the radius of the smaller cylinder?

5. Find the ratio of the circumference of the bases.

6. Find the ratio of the surface areas of the cylinders.

7. Find the ratio of the volumes of the cylinders.

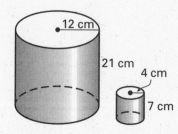

The solid is similar to a larger solid with the given scale factor. Find the surface area *S* and volume *V* of the larger solid.

8. Scale factor 1:2

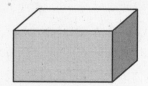

$S = 208$ in.^{2}

$V = 192$ in.^{3}

9. Scale factor: 1:3

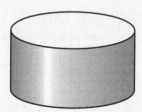

$S = 108\pi$ in.^{2}

$V = 108\pi$ in.^{3}

10. Scale factor: 1:4

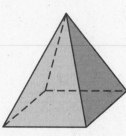

$S = 154$ cm^2

$V = 64$ cm^3

11. Scale factor 2:3

$S = 90\pi$ cm^2

$V = 100\pi$ cm^3

12. *Model Train* A toy model of a train is built with a scale of 1:12. If the model has a surface area of 94 square inches, what is the surface area of the actual train?

NAME _____ DATE _____

Practice B

For use with pages 766–772

Decide whether the solids are similar. If so, determine the scale factor.

1.

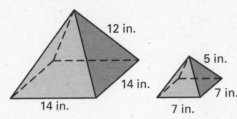

12 in.

14 in.

14 in.

5 in.

7 in.

7 in.

2.

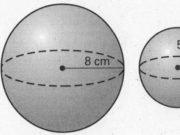

8 cm

5 cm

The solid is similar to a larger solid with the given scale factor. Find the surface area _S_ and volume _V_ of the larger solid.

3. Scale factor 1:2

$S = 208\pi$ in.^{2}

$V = 320\pi$ in.^{3}

4. Scale factor: 1:3

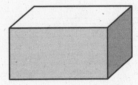

$S = 398$ in.^{2}

$V = 495$ in.^{3}

5. Scale factor: 2:3

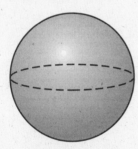

$S = 144\pi$ cm^2

$V = 288\pi$ cm^3

6. Scale factor 3:4

$S = 96\pi$ cm^2

$V = 96\pi$ cm^3

In Exercises 7–12, you and your friends decide to make a scale model of the water tower in your town.

7. You decide that 0.25 inch in your model will correspond to 12 inches of the actual water tower. What is the scale factor?

8. The top of the water tower has a diameter of 20 feet. Find the surface area of the top.

9. You decide to make the top of the water tower with silver foil. How many square inches of foil will you need?

10. The height of the actual water tower is 32 feet. What is the surface area of your scale model? Do not include the bottom base.

11. Find the volume of the actual water tower.

12. Use your result from Exercise 11 to find the volume of the scale model.

NAME _____ DATE _____

Practice C
For use with pages 766–772

Decide whether the solids are similar. If so, determine the scale factor.

1.

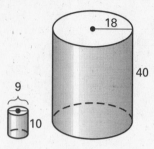

2.

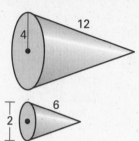

Complete the table.

		Surface Area	Volume	Scale Factor of A to B
3.	Solid A	64 in.²	28 in.³	1:2
	Solid B	_____	_____	
4.	Solid A	_____	_____	2:1
	Solid B	608 π in.²	1920 π in.³	
5.	Solid A	36 cm²	12 cm³	?
	Solid B	324 π in.²	?	
6.	Solid A	108 ft²	54 ft³	2:3
	Solid B	_____	_____	

Find the surface area and volume of the solid. Then use the scale factor to find the surface area and volume of the similar solid.

7. Scale factor 1:3

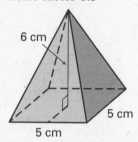

6 cm

5 cm

5 cm

8. Scale factor 1:4

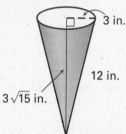

3 in.

12 in.

$3\sqrt{15}$ in.

9. Scale factor 2:7

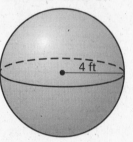

4 ft

In Exercises 10 and 11, use the following information.

You have purchased a scale model of a car. The scale factor is 1:24. The model is 2.9 inches high, 2.75 inches wide, and 6.4 inches long.

10. Find the dimensions of the car in feet.

11. If the rear cargo area of the actual car has a volume of 12.5 cubic feet, what is the volume of the rear cargo area of the model?

NAME _____ DATE _____

Reteaching with Practice

For use with pages 766–772

GOAL Find and use the scale factor of similar solids and use similar solids to solve problems

---

VOCABULARY

Two solids with equal ratios of corresponding linear measures, such as heights or radii, are called **similar solids**.

The common ratio of linear measures for a pair of similar solids is called the **scale factor** of one solid to the other solid.

Theorem 12.13 Similar Solids Theorem If two similar solids have a scale factor of $a:b$, then corresponding areas have a ratio of $a^2:b^2$, and corresponding volumes have a ratio of $a^3:b^3$.

EXAMPLE 1 *Identifying Similar Solids*

Decide whether the two solids are similar. If so, find the scale factor.

a.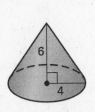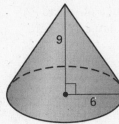

b.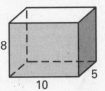

SOLUTION

a. The solids are similar because the ratios of corresponding linear measures are equal, as shown.

$$\text{radii: } \frac{4}{6} = \frac{2}{3} \qquad \text{heights: } \frac{6}{9} = \frac{2}{3}$$

The solids have a scale factor of 2:3.

b. The solids are not similar because the ratios of corresponding linear measures are not equal, as shown.

$$\text{widths: } \frac{4}{5} \qquad \text{lengths: } \frac{8}{10} = \frac{4}{5} \qquad \text{heights: } \frac{6}{8} = \frac{3}{4}$$

NAME ______________________ DATE __________

Reteaching with Practice

For use with pages 766–772

Exercises for Example 1

Decide whether the two solids are similar. If so, find the scale factor.

1. 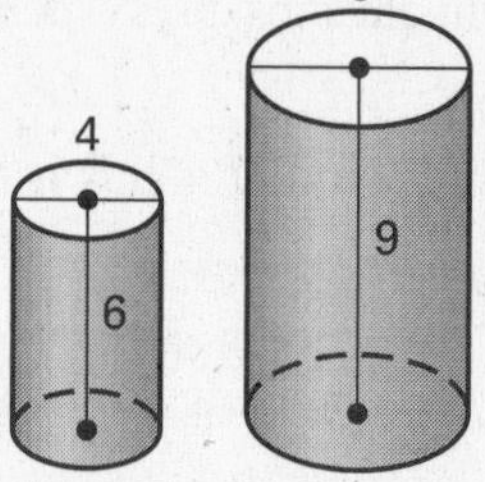

2. 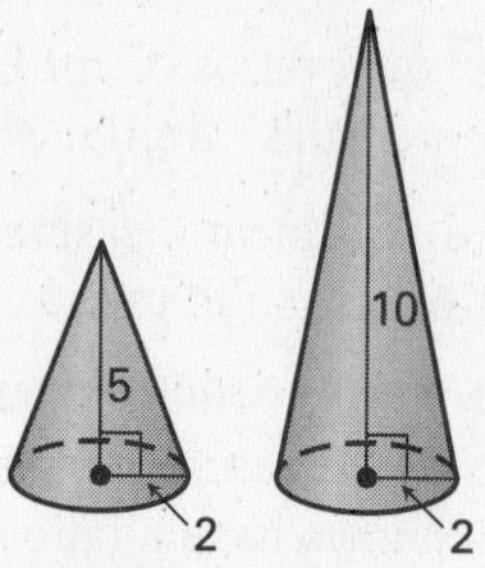

3. 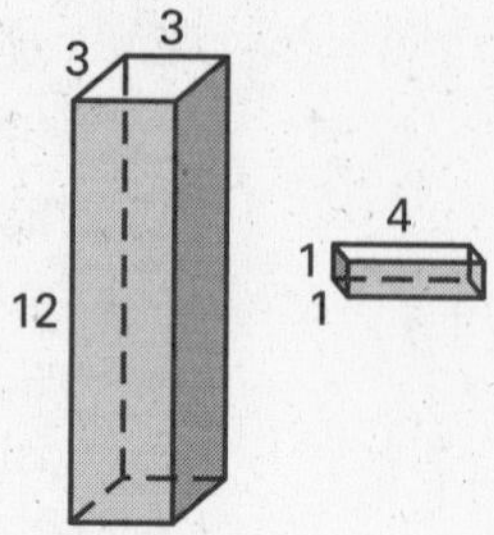

EXAMPLE 2 Using the Scale Factor of Similar Solids

The spheres are similar with a scale factor of 1:4. Find the surface area and volume of sphere B given that the surface area of sphere A is 144π square inches and the volume of sphere A is 288π cubic inches.

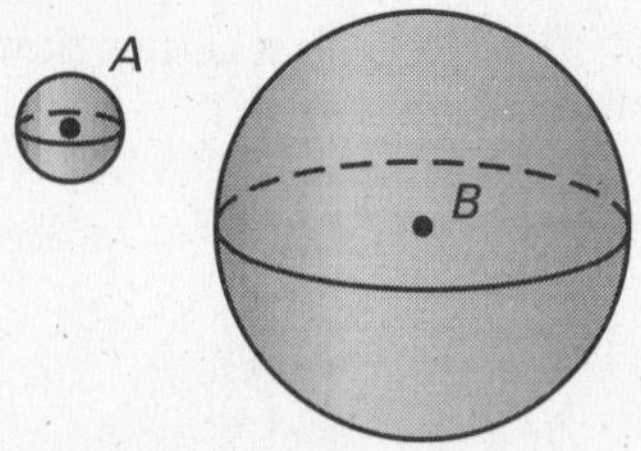

SOLUTION

Begin by using Theorem 12.13 to set up two proportions.

$$\frac{\text{Surface area of } A}{\text{Surface area of } B} = \frac{a^2}{b^2} \qquad \frac{\text{Volume of } A}{\text{Volume of } B} = \frac{a^3}{b^3}$$

$$\frac{144\pi}{\text{Surface area of } B} = \frac{1^2}{4^2} \qquad \frac{288\pi}{\text{Volume of } B} = \frac{1^3}{4^3}$$

$$\text{Surface area of } B = 2304\pi \qquad \text{Volume of } B = 18{,}432\pi$$

So, the surface area of sphere B is 2304π square inches and the volume of sphere B is $18{,}432\pi$ cubic inches.

Exercises for Example 2

The solid described is similar to a larger solid with the given scale factor. Find the surface area S and volume V of the larger solid.

4. A right cylinder with a surface area of 48π square centimeters and a volume of 45π cubic centimeters; scale factor 2:3

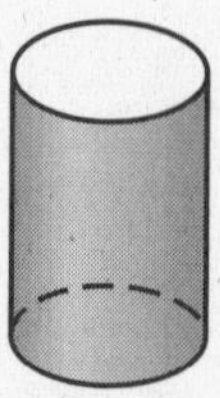

5. A right prism with a surface area of 82 square feet and a volume of 42 cubic feet; scale factor 1:2

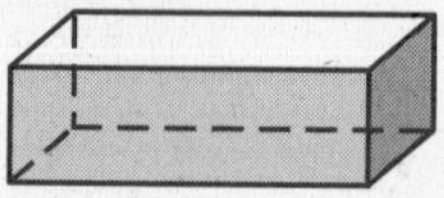

NAME _____ DATE _____

Quick Catch-Up for Absent Students

For use with pages 766–772

The items checked below were covered in class on (date missed) _____

Lesson 12.7: Comparing Similar Solids

_____ **Goal 1:** Find and use the scale factor of similar solids. (pp. 766–767)

Material Covered:

_____ Example 1: Identifying Similar Solids

_____ Student Help: Look Back

_____ Example 2: Using the Scale Factor of Similar Solids

_____ Example 3: Finding the Scale Factor of Similar Solids

Vocabulary:

similar solids, p. 766

_____ **Goal 2:** Use similar solids to solve real-life problems. (p. 768)

Material Covered:

_____ Example 4: Using Volumes of Similar Solids

_____ Student Help: Study Tip

_____ Example 5: Comparing Similar Solids

_____ Other (specify) _____

_____

Homework and Additional Learning Support

_____ Textbook (specify) _pp. 769–772_ _____

_____

_____ *Reteaching with Practice* worksheet (specify exercises)_____

_____ *Personal Student Tutor* for Lesson 12.7

Cooperative Learning Activity

GOAL **To investigate the relationship of surface area and volume in similar solids**

Materials: shoe box, construction paper, ruler, pencil, tape

Exploring Similar Solids

Two solids with equal ratios of corresponding linear measures are called similar solids. This ratio, called the scale factor, can be used to determine the corresponding areas and volumes of two similar solids.

Instructions

1 Close the shoe box and tape the lid onto the top of the box.

2 Measure the volume and the surface area of the shoe box.

3 Using a scale factor of 3:1, construct a smaller similar solid using the construction paper and tape.

4 Measure the surface area and volume of the constructed box.

Analyzing the Results

1. What is the surface area of the constructed box (in cm^2)?

2. What is the volume of the constructed box (in cm^3)?

3. What is the ratio of the surface areas of the boxes?

4. What is the ratio of the volumes of the boxes?

Geometry
Chapter 12 Resource Book

Interdisciplinary Application

For use with pages 766–772

Architectural Design

INDUSTRIAL ARTS The construction of scale models by architectural firms is extremely helpful to their clients. The models can be used to advertise a product such as individual homes. They are also used as aids in expansion plans for corporations such as hospitals and factories. Many clients of architectural firms comment on the business and aesthetic value of this important marketing tool.

For example, a contractor/developer builds a subdivision and the architects that designed the houses supply scale models of each style of house that was designed. Instead of simply seeing blueprints, sketches and floor plans of the house, a prospective buyer can see a three dimensional model.

In Exercises 1–4, use the information above.

1. What is the scale factor if the height of the model's walls are 8 inches tall and the height of the house's walls are 16 feet?

2. The outside dimensions of the model are 15 inches × 20 inches. What are the dimensions of the house in feet?

3. If the area of the living room in the model is 39 square inches, then what is the area of the house's living room in square feet?

4. A storage room in the model has a volume of 20 cubic inches. What is the volume of the storage room in the house in cubic feet?

NAME _____ DATE _____

Challenge: Skills and Applications

For use with pages 766–772

1. Suppose sand is being poured onto a cone-shaped pile, beginning at time $t = 0$, at the rate of 29.4 cubic inches per minute. At $t = 2$ minutes, the resulting cone has a diameter of 7 in. As the sand continues to be poured, the cone is always similar to its original shape, but it grows in size. Find the height and surface area of the cone at $t = 2$ minutes. (Remember to include the base of the pile.) Round your answers to the nearest hundredth.

In Exercises 2–5, use the information about a pair of similar solids to find all possible values of x.

2. Volume ratio is 8:27; surface area ratio is 12:x.

3. Surface area ratio is 16:25; volume ratio is x:125.

4. Scale factor is $3x$:$x + 5$; surface area ratio is $7x$:$x + 9$.

5. Scale factor is 1:2; volume ratio is $x - 2$:$x^2 - 1$.

In Exercises 6 and 7, use the following information.

The *density* of a material is the mass of the material per unit volume. The densities of some common materials are given at the right.

To find the mass of an object, multiply the volume by the density. For example, the mass of 4 cm³ of copper is 4 cm³ · 9.0 g/cm³ = 36 g.

To find the volume of an object, divide the mass by the density. For example, the volume of 386 g of gold is 386 g ÷ (19.3 g/cm³) = 20 cm³.

Material	Density
Glass	2.5 g/cm³
Aluminum	2.7 g/cm³
Iron	7.8 g/cm³
Copper	9.0 g/cm³
Silver	10.5 g/cm³
Gold	19.3 g/cm³

6. The scale factor of a silver figurine to a similar glass figurine is 2:5. If the mass of the silver figurine is 50.4 g, what is the mass of the glass figurine?

7. The mass of an iron frying pan is 499.2 g. A similar aluminum frying pan has mass 337.5 g. The surface area of the iron pan is 320 cm².

   a. Find the scale factor of the iron pan to the aluminum pan.

   b. What is the surface area of the aluminum pan?

NAME _____ DATE _____

Chapter Review Games and Activities

For use after Chapter 12

Determine the word or words described by the given phrase. Then unscramble the specified letters of your answers to spell out the name of a famous mathematician mentioned in the chapter.

1. A line segment formed by the intersection of two faces of a polyhedron. (first letter)

2. A polyhedron with eight faces. (first letter)

3. A segment from the center of a sphere to a point on the sphere. (first letter)

4. Has a circular base and a vertex that is not in the same plane as the base. (last letter)

5. A polyhedron with two congruent faces, called bases, that lie in parallel planes. (second letter)

6. A polyhedron with 12 faces. (first letter)

7. Two solids with equal ratios of corresponding linear measures, such as heights or radii. (sixth letter)

8. The number of cubic units contained in the interior of a solid. (fourth letter)

9. Consists of all the segments that connect the vertex of a cone with points on the base edge. (first letter)

10. Five regular polyhedra, named after the Greek mathematician and philosopher Plato. (sixth letter)

11. The locus of points in space that are a given distance from a point. (third letter)

12. If two solids have the same height and the same cross-sectional area at every level, then they have the same volume. (fifth letter)

13. A solid with congruent circular bases that lie in parallel planes. (seventh letter)

Famous Mathematician: _____ _____

NAME _____ DATE _____

Chapter Test A

Determine the number of faces, vertices, and edges of the solids.

1.
2.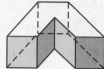
3.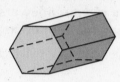

Answers

1. _____
2. _____
3. _____
4. _____
5. _____
6. _____
7. _____
8. _____
9. _____
10. _____
11. _____

Describe the shape formed by the intersection of the plane and the cube.

4.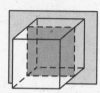

5.

Find the surface area of the right prism or right cylinder. Round the result to two decimal places.

6.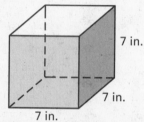
7 in.
7 in.
7 in.

7.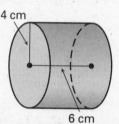
4 cm
6 cm

Find the area of a lateral face of the regular pyramid. Then find the surface area of the regular pyramid. Round the results to one decimal place.

8.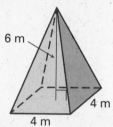
6 m
4 m
4 m

9.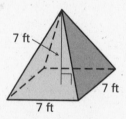
7 ft
7 ft
7 ft

Find the volume and surface area of the right cone. Round the result to one decimal place.

10.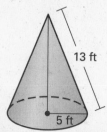
13 ft
5 ft

11.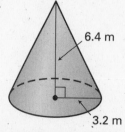
6.4 m
3.2 m

NAME _____ DATE _____

Chapter Test A

Find the volume of the right prism.

12.
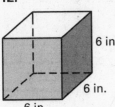

6 in.

6 in.

6 in.

13.
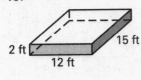

2 ft 12 ft 15 ft

14.
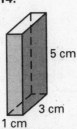

5 cm

3 cm

1 cm

Find the volume of the right cylinder. Round the result to two decimal places.

15.
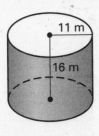

11 m

16 m

16.
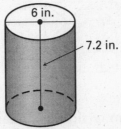

6 in.

7.2 in.

Find the volume of the pyramid. Each pyramid has a regular polygon for a base. Round the result to two decimal places.

17.
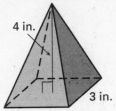

4 in.

3 in.

18.
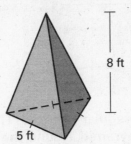

8 ft

5 ft

Find the volume and surface area of the sphere. Round the result to two decimal places.

19.
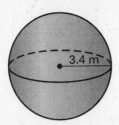

3.4 m

20.
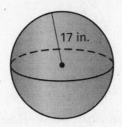

17 in.

Decide whether the solids are similar.

21.
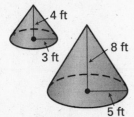

4 ft

3 ft

8 ft

5 ft

22.
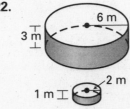

6 m

3 m

2 m

1 m

12.	_____
13.	_____
14.	_____
15.	_____
16.	_____
17.	_____
18.	_____
19.	_____
20.	_____
21.	_____
22.	_____

Review and Assess

NAME _____ DATE _____

Chapter Test B

For use after Chapter 12

Determine the number of faces, vertices, and edges of the solids.

1.
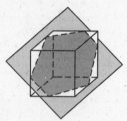

2.

3.
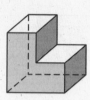

Describe the shape formed by the intersection of the plane and the cube.

4.
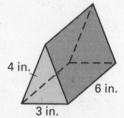

5.
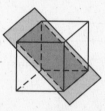

Find the surface area of the right prism or right cylinder. Round the result to two decimal places.

6.
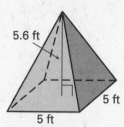

4 in.
6 in.
3 in.

7.
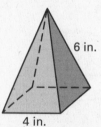

8.4 in.
8 in.

Find the area of a lateral face of the regular pyramid. Then find the surface area of the regular pyramid. Round the result to one decimal place.

8.
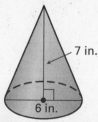

5.6 ft
5 ft
5 ft

9.
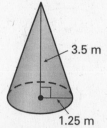

6 in.
4 in.

Find the volume and surface area of the right cone. Round the result to one decimal place.

10.

7 in.
6 in.

11.

3.5 m
1.25 m

Answers
1. _____
2. _____
3. _____
4. _____
5. _____
6. _____
7. _____
8. _____
9. _____
10. _____
11. _____

Find the volume of the solid.

12.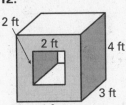

2 ft
2 ft
4 ft
3 ft
4 ft

13.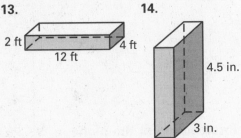

2 ft
12 ft
4 ft

14.

4.5 in.
3 in.
1 in.

12.	_____
13.	_____
14.	_____
15.	_____
16.	_____
17.	_____
18.	_____
19.	_____
20.	_____
21.	_____
22.	_____

Find the volume of the solid. Round the result to two decimal places.

15.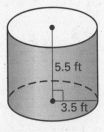

5.5 ft
3.5 ft

16.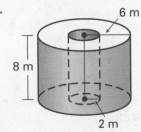

6 m
8 m
2 m

Find the volume of the pyramid. Each pyramid has a regular polygon for a base. Round the result to two decimal places.

17.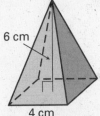

6 cm
4 cm

18.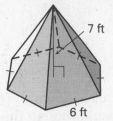

7 ft
6 ft

Find the volume and surface area of the sphere. Round the result to two decimal places.

19.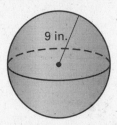

9 in.

20.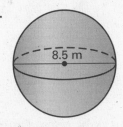

8.5 m

Decide whether the solids are similar.

21.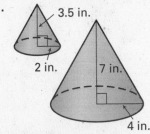

3.5 in.
2 in.
7 in.
4 in.

22.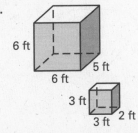

6 ft
6 ft
5 ft
3 ft
3 ft
2 ft

Review and Assess

Determine the number of faces, vertices, and edges of the solids.

1.

2.

3.

Describe the shape formed by the intersection of the plane and the cube.

4.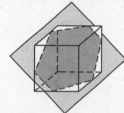

5.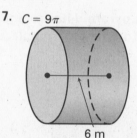

Answers

1. _____
2. _____
3. _____
4. _____
5. _____
6. _____
7. _____
8. _____
9. _____
10. _____
11. _____

Find the surface area of the right prism or right cylinder. Round the result to two decimal places.

6.

5 cm
7.2 cm
6 cm

7. $C = 9\pi$

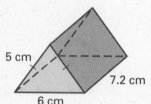

6 m

Find the area of a lateral face of the regular pyramid. Then find the surface area of the regular pyramid. Round the results to one decimal place.

8.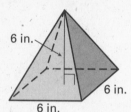

6 in.
6 in.
6 in.

9.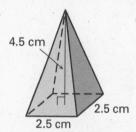

4.5 cm
2.5 cm
2.5 cm

Find the volume and surface area of the right cone. Round the result to one decimal place.

10.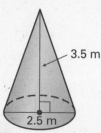

3.5 m
2.5 m

11.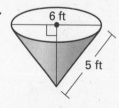

6 ft
5 ft

Review and Assess

NAME _____ DATE _____

Chapter Test C

For use after Chapter 12

Find the volume of the solid.

12.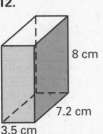

8 cm

7.2 cm

3.5 cm

13.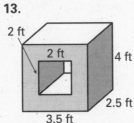

2 ft

2 ft

4 ft

2.5 ft

3.5 ft

14.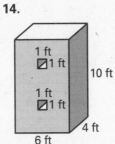

1 ft
1 ft

10 ft

1 ft
1 ft

4 ft

6 ft

**Find the volume of the solid. Round the result to two
decimal places.**

15.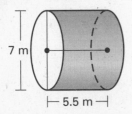

7 m

5.5 m

16.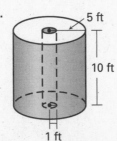

5 ft

10 ft

1 ft

**Find the volume of the pyramid. Each pyramid has a regular
polygon for a base. Round the result to two decimal places.**

17.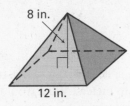

8 in.

12 in.

18.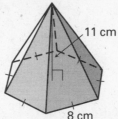

11 cm

8 cm

**Find the volume and surface area of the sphere. Round the
result to two decimal places.**

19.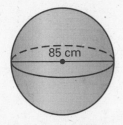

85 cm

20.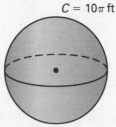

$C = 10\pi$ ft

Decide whether the solids are similar.

21.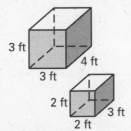

3 ft

4 ft

3 ft

2 ft

3 ft

2 ft

22.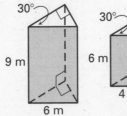

30°

30°

9 m

6 m

6 m

4 m

12.	_____
13.	_____
14.	_____
15.	_____
16.	_____
17.	_____
18.	_____
19.	_____
20.	_____
21.	_____
22.	_____

Review and Assess

NAME _____ DATE _____

SAT/ACT Chapter Test

For use after Chapter 12

1. Which of the figures shown below are not a polyhedron?

I

II

III

    (A) I only
    (B) II only
    (C) III
    (D) I and III
    (E) II and III

2. A polyhedron has 14 faces and 20 edges. How many vertices does it have?

  (A) 6   (B) 8   (C) 10

  (D) 12   (E) 14

3. What is the surface area of the right prism shown below?

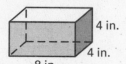
4 in.
4 in.
8 in.

  (A) 48 in.^{2}
  (B) 64 in.^{2}
  (C) 128 in.^{2}
  (D) 144 in.^{2}
  (E) 160 in.^{2}

4. What is the surface area of the right cylinder shown below?

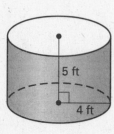
5 ft
4 ft

  (A) about 226 ft^2
  (B) about 157 ft^2
  (C) about 257 ft^2
  (D) about 283 ft^2
  (E) about 151 ft^2

5. What is the radius of a sphere whose volume is 330 cubic meters?

  (A) 2.79 m   (B) 3.91 m

  (C) 4.29 m   (D) 5.14 m

  (E) 6.21 m

6. What is the surface area of the right cone shown below?

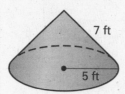
7 ft
5 ft

  (A) about 241.9 ft^2
  (B) about 110 ft^2
  (C) about 204.2 ft^2
  (D) about 188.5 ft^2
  (E) about 152.9 ft^2

7. The volume of a cube is 125 cubic inches. What is the surface area of the cube?

  (A) 25 in.^{2} (B) 150 in.^{2}

  (C) 100 in.^{2} (D) 250 in.^{2}

  (E) 625 in.^{2}

8. The volume of the rectangular prism shown below is 4. What is the value of x?

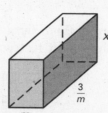
x
$\frac{3}{m}$
m

  (A) $\frac{m}{4}$ (B) $4m$

  (C) $\frac{3}{4}$ (D) $\frac{m}{3}$

  (E) $\frac{4}{3}$

9. The surface area of a sphere with a radius of 3 meters is 36π square meters. What would the surface area be after the radius is doubled?

  (A) 144π m^2 (B) 162π m^2

  (C) 72π m^2 (D) 81π m^2

  (E) 64π m^2

10. The dimensions of the right cube prism shown below are doubled. How many times larger is the volume of the new prism?

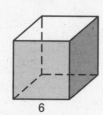
6

  (A) 2
  (B) 6
  (C) 8
  (D) 10
  (E) 4

NAME _____ DATE _____

Alternative Assessment and Math Journal

For use after Chapter 12

JOURNAL **1.** Explain how Euler's Theorem relates faces, vertices, and edges
of a polyhedron. Use Euler's Theorem to complete the following.

    a. Faces: 7   **b.** Faces: 12   **c.** Faces: __?__

       Vertices: 7   Vertices: __?__   Vertices: 19

       Edges: __?__   Edges: 18   Edges: 36

**MULTI-STEP
PROBLEM** **2.** You work in the packaging department of a company that makes
soup. The old soup can is 4 inches high and has a diameter of
3 inches. Your department is asked to design a new can that will
hold the same amount of soup but the new can should be 1 inch
taller.

    a. Find the volume of the old soup can.

    b. Explain what you think will happen to the diameter of the soup
can when its height is increased.

    c. Find the diameter of the new soup can.

    d. Your department is asked to design a bigger soup can that will
hold one-and-a-half times more soup than the old can. Using the
dimensions of the old soup can, the diameter is to stay the same.
Find the height of the bigger soup can.

3. *Critical Thinking* As the company is switching from the old soup
can to the new soup can, there is concern about the label that goes
around the can. You are asked if the new label will use the same
amount of paper as the old label. Remember the label on the can
does not cover the top or bottom of the can. Find the amount of
paper used in the label for the old and the new soup can. Are they
the same? If not, which can uses more?

4. *Writing* Write a paragraph explaining if the old soup can and the
new soup can are similar; and if the old soup can and the bigger soup
can are similar. Be sure to include diagrams.

Alternative Assessment Rubric

For use after Chapter 12

JOURNAL SOLUTION

1. Complete answers should include:

- The number of faces (F), vertices (V), and edges (E) of a polyhedron are related by the formula $F + V = E + 2$.

   a. Edges $= 12$

   b. Vertices $= 8$

   c. Faces $= 19$

MULTI-STEP PROBLEM SOLUTION

2. a. $V = 9\pi \approx 28.27$ cubic inches

   b. Answers may vary. Answers should suggest the diameter decreasing.

   c. $d \approx 2.68$ inches

   d. $h \approx 6$ inches

3. The old can uses about 37.70 square inches. The new can uses about 47.12 square inches. No, they do not use the same amount of paper. The new can will use more.

4. Answers may vary. Answers should show that none of the cans are similar. Diagrams should be included.

MULTI-STEP PROBLEM RUBRIC

4 Students answer all parts of the problem correctly. Clear explanations are given for the solutions. Students explain Euler's Theorem correctly. Students show they know the meaning of similar figures. Students' diagrams are correctly labeled.

3 Students complete all questions. Students' work may have a minor mathematical error. Explanations are good. Students explain Euler's Theorem. Students have a general idea of similar figures. Students' diagrams may have an incorrect label.

2 Students complete all questions. Students' work has several mathematical errors. Explanations are sufficient. Students explain Euler's Theorem. Students' explanation of similar figures is not clear. Students' diagrams are incorrect.

1 Students do not answer all questions completely. Students' explanations are incorrect. Students do not explain Euler's Theorem correctly. Students' explanation of similar figures is incomplete. Students' diagrams are incomplete.

Review and Assess

NAME _____ DATE _____

Project: *Spherical Modeling*

For use with Chapter 12

OBJECTIVE **Design and create spherical models.**

MATERIALS cardboard or paper, compass, protractor, ruler, scissors, tape, glue, plastic bags

INVESTIGATION *Making a Spherical Cuboctahedron* Your model
will be constructed by gluing together eight
"spherical triangles" and six "spherical squares."
The patterns are shown below. Note that you will cut
out only the thin curved strip between the two solid
arcs. The dashed lines and angle measures are
included so you can create your templates. As
shown at the right, the two half-sector sections at the
ends of a strip are taped together to make one edge.

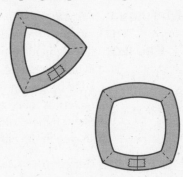

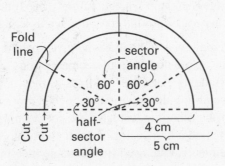

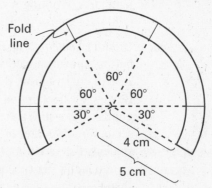

Pattern for spherical triangle   **Pattern for spherical square**

Step 1 Create one template from each pattern. Then use the template to trace the
strips. You will need 8 strips for triangles and 6 strips for squares.

Step 2 Cut out all of the strips. Crease each on the solid fold lines, and tape together
the two half-sector sections.

Step 3 Glue the spherical shapes together so that each triangle is bordered by three
squares and each square is bordered by four triangles.

Making Other Spherical Polyhedra Select and create at least one model from this
table. Create a curved strip pattern similar to the ones above, using the same radii
measurements as shown. Exact sector angle measures are given, but you may need
to use approximate measures. Note that each model uses only one type of shape.

Polyhedron name	Number of strips needed	Number of sectors on a strip	Sector angle measure	Number of shapes meeting at a vertex
Tetrahedron	4	3 (2 full & 2 half)	109°28'	3
Cube	6	4 (3 full & 2 half)	70°32'	3
Octahedron	8	3 (2 full & 2 half)	90°	4

PRESENT YOUR RESULTS Show and identify the two spherical models you created. Discuss the number of
edges, vertices, and faces in each model. Show and explain your rationale and
calculations. Does Euler's Theorem still hold in spherical geometry? Also include
a description of how the models you created are alike and how they are different.

Project: Teacher's Notes

For use with Chapter 12

GOALS
- Create spherical models and compare their characteristics.
- Count faces, edges, and vertices to verify Euler's Theorem for spherical polyhedra.

MANAGING THE PROJECT

Classroom Management   You may want students to cut out the strips for all the models they will make, and then do all the taping and gluing at one time. A plastic bag is ideal for keeping the strips together prior to gluing. (If desired, you can avoid the taping step. Instead, two half-sector sections can be joined to form an edge by gluing them both to a single full section of a bordering strip.) Using stiff materials makes the templates easier to trace around and helps the models hold their shape. Colors can be used, with different colors chosen for the two different shapes in the cuboctahedron. Models can be displayed in a class mobile.

Cooperative Learning   Suggest that the students have someone check their templates. Working with a partner can help when creating strips (one person can hold the template while the other traces around it) and when assembling a model.

Guiding Students' Work   You may wish to analyze the given patterns together before students create templates. Note that the half-sector sections are used for ease in construction. Be sure students mark the fold lines as they trace each strip. Review the description for assembling the spherical cuboctahedron prior to gluing. Make sure patterns for the other models have the correct number of full sectors in the center of the strip and a half sector on each end of it. Assure students that close approximations may be used for the angle measures. Students should realize that the number of sectors for a curved strip indicates the spherical shape: 3 sectors for a triangle and 4 for a square. Students need to keep in mind the number of shapes meeting at a vertex; it may be helpful to arrange bordering shapes before gluing.

RUBRIC   **The following rubric can be used to assess student work.**

4   The student neatly and accurately assembled two spherical models. The student used effective strategies, correctly found the number of vertices, edges, and faces for each model, and concluded that Euler's Theorem holds in spherical geometry. A thoughtful comparison of the three models is included.

3   The student assembled two spherical models. The student correctly found the number of vertices, edges, and faces for each model and concluded that Euler's Theorem holds, but did not give good strategies for counting. The student gave a sketchy comparison of the two models built.

2   The student poorly assembled two spherical models. Due to some errors in finding the number of vertices, edges, and faces, the student failed to realize that Euler's Theorem still holds. The student didn't give an accurate comparison of the two models built.

1   The student did not assemble a second spherical model. The construction of the spherical cuboctahedron was sloppy or incorrect. Calculations for the number of vertices, edges, and faces are missing or inaccurate. The student neither recognized that Euler's Theorem still holds nor gave a comparison of two models.

NAME _____ DATE _____

Cumulative Review

For use after Chapters 1–12

State the postulate or theorem you would use to prove the triangles are congruent. (4.3, 4.4, 4.6)

1.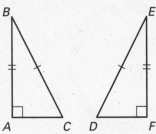

2.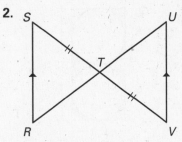

List the sides of the triangle in order from shortest to longest. (5.5)

3.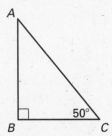

4.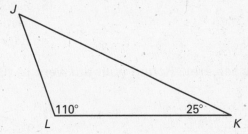

Find the value of x. (6.4)

5.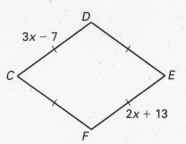

6.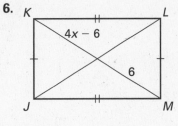

The two polygons are similar. Find the values of x and y. (8.3)

7.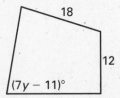

8.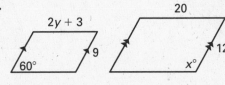

Find the values of x and y to the nearest tenth. (9.5)

9.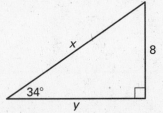

10.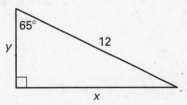

Review and Assess

Find the area of the shaded region. Round your answers to two decimal places. (11.5)

11.

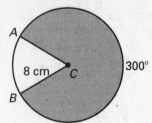

12.

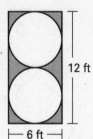

13.

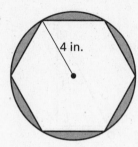

Use Euler's Theorem to find the unknown number. (12.1)

14. Faces: 8
Vertices: _____?_____
Edges: 12

15. Faces: 5
Vertices: 9
Edges: _____?_____

Find the surface area. Round your answers to two decimal places. (12.2, 12.3)

16.

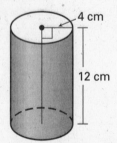

17.

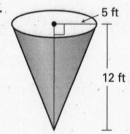

Find the volume. (12.4, 12.5)

18.

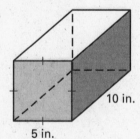

19.

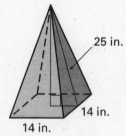

Find the volume and surface area of the sphere described. Round your answers to two decimal places. (12.6)

20. A sphere with a radius of 4 inches

21. A sphere with a diameter of 1.5 feet

ANSWERS

Chapter Support

Parent Guide
12.1: 7 faces; hexagon   **12.2:** ,888 cm^2
12.3: about 678.6 in.^{2} **12.4:** 282.7 m^3
12.5: 147,000 ft^3 **12.6:** 201,061,929 mi^2;
about 32%   **12.7:** 3 : 4
Home Involvement Activity: An 8 by 8 by 2 box
has the maximum volume.

Prerequisite Skills Review
1. $\frac{2}{1}$ **2.** $\frac{4}{3}$ **3.** $\frac{2}{3}$ **4.** 32 in.^{2} **5.** 24 ft^2
6. 120 ft^2 **7.** 21.2 cm^2 **8.** 81.0 ft^2
9. 259.8 in.^{2}

Strategies for Reading Mathematics
1. *ABCD, DCGH, CBFG*   **2.** *EFGH, ABFE,*
DAEH   **3.** $\overline{AB}, \overline{BC}, \overline{CD}, \overline{DA}, \overline{BF}, \overline{FG},$
$\overline{CG}, \overline{HG}, \overline{DH}$; they are solid; $\overline{AE}, \overline{EF}, \overline{EH}$; they are
dashed.   **4.** front: $\triangle MPQ, \triangle MNP$; back and
bottom: $\triangle MNQ, \triangle QNP$

Lesson 12.1

Warm-Up Exercises
1. 9   **2.** 6   **3.** 10   **4.** 12   **5.** n

Daily Homework Quiz
1. $\frac{2}{7}$, 0.29, or 29%   **2.** $\frac{1}{2}$ or 50%   **3.** 20

Lesson Opener
Allow 20 minutes.
1. 4 equilateral triangles   **2.** 6 squares
3. 8 equilateral triangles   **4.** 12 regular
pentagons   **5.** 20 equilateral triangles

Practice A
1. No; some of the faces are not polygons.
2. Yes; the figure is a solid that is bounded by
polygons that enclose a single region of space.
3. Yes; the figure is a solid that is bounded by
polygons that enclose a single region of space.

4. $F = 6, V = 8, E = 12$ **5.** $F = 7, V = 10$,
$E = 15$ **6.** $F = 5, V = 6, E = 9$
7. no; all of its faces are not $\cong$; convex   **8.** yes;
all of its faces are $\cong$ regular polygons; convex
9. no; all of its faces are not $\cong$; not convex
10. 8   **11.** 12   **12.** 18
13. yes; $8 + 12 = 18 + 2$ **14.** 5   **15.** 6
16. 9   **17.** yes; $5 + 6 = 9 + 2$
18. 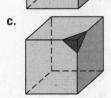   **19.**

Practice B
1. No; some of the faces are not polygons.
2. Yes; the figure is a solid that is bounded by
polygons that enclose a single region of space.
3. Yes; the figure is a solid that is bounded by
polygons that enclose a single region of space.
4. $F = 6, V = 8, E = 12$ **5.** $F = 8, V = 12$,
$E = 18$ **6.** $F = 6, V = 6, E = 10$
7. yes; all of its faces are $\cong$ regular polygons;
convex   **8.** no; not all of its faces are $\cong$;
not convex   **9.** no; not all of its faces are regular
polygons; convex   **10.** 8   **11.** 5   **12.** 15
13. rectangle   **14.** regular hexagon   **15.** triangle
16. a.   **b.**

c.   **d.**

Practice C
1. No; some of the faces are not polygons.
2. Yes; the figure is a solid that is bounded by
polygons that enclose a single region of space.

Lesson 12.1 continued

3. Yes; the figure is a solid that is bounded by polygons that enclose a single region of space.

4. $F = 7$, $V = 10$, $E = 15$; $7 + 10 = 15 + 2$

5. $F = 10$, $V = 16$, $E = 24$; $10 + 16 = 24 + 2$

6. $F = 7$, $V = 7$, $E = 12$; $7 + 7 = 12 + 2$

7. false 8. true 9. false 10. true

11. false 12. false 13. rectangle

14. rectangle 15. rectangle 16. 20 17. 24

18. 24

Reteaching with Practice

1. yes; all of its faces are polygons which form a solid enclosing a single region of space.

2. yes; all of its faces are polygons which form a solid enclosing a single region of space.

3. no; some of its faces are not polygons.

4. 5 faces, 5 vertices, 8 edges

5. 6 faces, 8 vertices, 12 edges

6. 7 faces, 10 vertices, 15 edges 7. 10

Interdisciplinary Application

1. No; a rhombus is not a regular polygon; all sides are congruent in a rhombus, but not all angles. 2. 24 3. No; unless the triangles are equilateral triangles, it cannot be a regular polyhedron. 4. 26 5. octagon

Challenge: Skills and Applications

1. $F = 12$, $V = 16$, $E = 28$; 0 2. $F = 18$, $V = 24$, $E = 42$; 0 3. $F = 13$, $V = 18$, $E = 33$; -2 4. $F = 20$, $V = 32$, $E = 56$; -4

5. $F = 28$, $V = 44$, $E = 74$; -2 6. $F = 24$, $V = 40$, $E = 70$; -6 7. 17 8. 20 9. 16

10. 60 11. 4

Lesson 12.2

Warm-Up Exercises

1. 54 ft^2 2. 150 cm^2 3. 256 in.^{2} 4. 56.1 m^2

Daily Homework Quiz

1. False; a cylinder is not a polyhedron because its faces are not polygons. 2. trapezoid

3. 7, 15, 10; $7 + 10 = 15 + 2$ 4. 24

Lesson Opener

Allow 15 minutes.

1. 2 equilateral triangles, 3 rectangles

2. 2 squares, 4 rectangles

3. 2 regular pentagons, 5 rectangles

4.

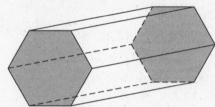

2 regular hexagons, 6 rectangles

Practice A

1. right pentagonal prism 2. regular pentagon

3. rectangle 4. 5 5. *Sample answer:* $\overline{AF}$, $\overline{BG}$, $\overline{CH}$ 6. right circular cylinder 7. circle

8. $\overline{AC}$ 9. $\overline{AB}$ 10. cube 11. cylinder

12. triangular prism 13. 432 in.^{2} 14. 72 cm^2

15. 48 in.^{2} 16. 376.99 cm^2 17. 408.41 in.^{2}

18. 967.61 in.^{2}

Practice B

1. right hexagonal prism 2. regular hexagon

3. rectangle 4. 6 5. *Sample answer:* $\overline{AG}$, $\overline{BH}$, $\overline{CI}$ 6. cylinder 7. cube 8. triangular prism

9. 392 cm^2 10. 180 in.^{2} 11. 233.67 cm^2

12. 282.74 ft^2 13. 359.08 cm^2

14. 1470.27 in.^{2} 15. 6 m 16. 3.0 cm

17. 14.0 in.

Practice C

1. right circular cylinder 2. right rectangular prism 3. right hexagonal prism 4. right circular cylinder 5. cylinder 6. pentagonal prism

7. hexagonal prism 8. 39.06 in.^{2}

9. 607.21 cm^2 10. 7118.09 cm^2 11. 8.5 m

12. 11.0 cm 13. 16.0 in.

Reteaching with Practice

1. 235.6 m^2 2. 472 ft^2 3. 141.7 in.^{2}

4. 326.7 m^2 5. 402.1 cm^2 6. 254.5 ft^2

Lesson 12.2 continued

Real-Life Application

1. 80 square feet   2. less than one gallon

3. ≈ 28.28 ft^2 4. one square   5. yes

Challenge: Skills and Applications

1. a.   2. a.

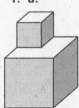   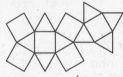

2b. $54 + 18\sqrt{3} \approx 85.18$ cm^2

1b. 252 in.^{2}

3. 722 m^2 4. 1040 ft^2 5. 354 cm^2 6. 8 in.

7. 9 mm   8. 7 ft

9. a. 7 in.   b.   5 in.

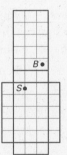

Lesson 12.3

Warm-Up Exercises

1. 112 cm^2 2. $170\pi \approx 534.07$ ft^2 3. 274 m^2

4. $22.5\pi \approx 70.69$ in.^{2}

Daily Homework Quiz

1. 8, 8   2. 213.57 cm^2 3. 477.52 in.^{2} 4. 4

Lesson Opener

Allow 5 minutes.

1. Check sketches. Answers will vary. *Sample answer:* A tipi is technically a pyramid, but it seems more like a cone because the use of so many poles makes it quite circular.

Technology Activity

1. no   2. Radius; the surface area increases more quickly for each unit of change in the radius than for each unit of change in the slant height.

Practice A

1. square pyramid   2. square   3. triangle

4. 4   5. *Sample answer:* $\overline{AB}, \overline{AC}, \overline{AD}$

6. $4\sqrt{10}$ 7. $\sqrt{265}$ 8. 144 cm^2 9. 144 in.^{2}

10. ≈ 63.33 m^2 11. 250π cm^2 12. 33π in.^{2}

13. 90π cm^2

14. 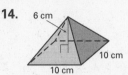 $S = 256.2$ cm^2

Practice B

1. $3\sqrt{29}$ 2. $\sqrt{349}$ 3. $\sqrt{65}$

4. ≈ 103.47 cm^2 5. ≈ 299.93 cm^2

6. ≈ 302.67 cm^2 7. 384π in.^{2} 8. 88.65π in.^{2}

9. 126.75π cm^2

10. 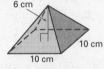   11.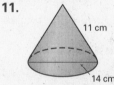

$S = 256.2$ cm^2

$S = 395.8$ cm^2

12. 75 ft^2 13. 203.6 cm^2 14. 284.4 in.^{2}

Practice C

1. $\sqrt{109}$ 2. $2\sqrt{15}$ 3. $3\sqrt{10}$

4. ≈ 27.71 in.^{2} 5. ≈ 245.02 cm^2

6. ≈ 210.19 ft^2 7. 1080π mm^2

8. 147π in.^{2} 9. $\approx 157.27\pi$ cm^2

10. $p = 7$ cm, $q = \sqrt{149}$ cm   11. $l \approx 9$ in., $h \approx \sqrt{56}$ in.   12. $l \approx 10$ cm

Reteaching with Practice

1. 16 cm^2 2. 105.2 m^2 3. 323.8 in.^{2}

4. 175.9 in.^{2} 5. 282.7 m^2 6. 267 ft^2

Interdisciplinary Application

1. ≈ 131 ft   2. $\approx 61,800$ ft^2 3. a. ≈ 28 ft

b. ≈ 20 ft   4. a. 2128 ft^2 b. 60,456 ft^2

Math and History Application

1. Cube: $6 + 8 = 12 + 2$;
Tetrahedron: $4 + 4 = 6 + 2$ 2. $5\sqrt{3}s^2$

Lesson 12.3 *continued*

Challenge: Skills and Applications

1. a. $TV = 29$ in., $SV \approx 35.81$ in.

b.

c. 1470 in.^{2}

2. 377.53 cm^2 **3.** 194.64 ft^2 **4. a.** 4.25 cm
b. $h \approx 3.78$ cm; $r \approx 1.95$ cm
5. a. 116.76°; 37 in.   **b.** 1847.26 in.^{2}

Quiz 1

1. not regular, convex; 6 vertices

2. regular, convex; 4 vertices

3. not regular, convex; 8 vertices

4. 1130.97 in.^{2} **5.** 216 m^2 **6.** 275.10 cm^2

Lesson 12.4

Warm-Up Exercises

1. 16 units^2 **2.** 49π, or about 153.9 cm^2
3. $4\sqrt{3}$, or about 6.93 m^2
4. $6\sqrt{3}$, or about 10.4 ft^2

Daily Homework Quiz

1. $36 + 12\sqrt{73} \approx 138.53$ ft^2
2. $3\sqrt{34} \approx 17.49$ mm   **3.** 216π in.^{2}
4. $x = 10$ m, $l \approx 12.11$ m

Lesson Opener

Allow 10 minutes.

1. *Sample answer:*

	Volume of box (cubic cm)	Net wt. of cereal (grams)	Volume: Net wt.
Small	3078	396	≈ 7.8
Medium	4410	567	≈ 7.8
Large	$\approx 11,401$	1389	≈ 8.2

The net weights are proportional to the volumes for the small and medium boxes, but not for the large box, as shown by the ratios in the table. They should be proportional if the boxes are equally full.

The large box has more wasted space; it should be able to contain more cereal to have the same ratio as the others.

2. *Sample answer:*

	Volume of can (cubic cm)	Net wt. of chili (grams)	Volume: Net wt.
Small	≈ 436	425	≈ 1.0
Large	≈ 1116	1130	≈ 1.0
Jumbo	≈ 3085	3062	≈ 1.0

Yes, the net weights are proportional to the volumes, as shown by the equal ratios in the table. This should be the case if all the cans are full.

Practice A

1. 72 unit cubes; 3 layers of 12 rows of 2 cubes each   **2.** 378 unit cubes; 9 layers of 7 rows of 6 cubes each   **3.** 384 square units; 6 layers of 8 rows of 8 cubes each   **4.** 64 in.^{3} **5.** 260 cm^3
6. 12 ft^3 **7.** $36\sqrt{3} \approx 62.35$ cm^3 **8.** 144 in.^{3}
9. $192\sqrt{3} \approx 332.55$ in.^{3} **10.** 565.49 cm^3
11. 552.92 in.^{3} **12.** 832.50 cm^3 **13.** 4800 ft^3
14. 3150 ft^3

Practice B

1. 240 cm^3 **2.** 520 ft^3 **3.** 96 cm^3
4. $81\sqrt{3} \approx 140.30$ in.^{3} **5.** ≈ 636.53 cm^3
6. 540 in.^{3} **7.** 2061.20 cm^2 **8.** 339.29 ft^3
9. 477.52 cm^3 **10.** 7 cm   **11.** ≈ 12 ft   **12.** 6.5 in.

13.

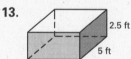

2.5 ft
5 ft
5 ft

$V = 62.5$ ft^3

14.

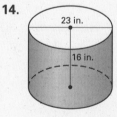

23 in.
16 in.

$V = 6647.61$ in.^{3}

15. Each pillar requires about 124.71 in.^{3}. All four require about 498.84 in.^{3}

Practice C

1. 468 in.^{3} **2.** ≈ 263.06 in.^{3} **3.** 1533 cm^3
4. ≈ 3562.57 in.^{3} **5.** ≈ 2429.27 cm^3
6. ≈ 326.73 in.^{3} **7.** 216 cm^3 **8.** 90 in.^{3}

Lesson 12.4 *continued*

9. $16\pi \approx 50.3 \text{ m}^3$

10. $15{,}625\pi \approx 49{,}087.4 \text{ in.}^3$

11. $\approx 2.4 \text{ cm}$ **12.** $\approx 2.9 \text{ ft}$ **13.** $\approx 4 \text{ in.}$

14. 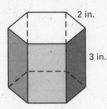 **15.**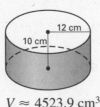

$V \approx 31.2 \text{ in.}^3$ $V \approx 4523.9 \text{ cm}^3$

16. 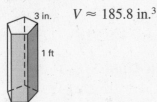 $V \approx 185.8 \text{ in.}^3$

Reteaching with Practice

1. 216 cm^3 **2.** 84.8 in.^3 **3.** 160 in.^3

4. 3 **5.** 4 **6.** 2.5

Real-Life Application

1. $\approx 1.57 \text{ in.}^3$ **2.** $\approx 18.84 \text{ in.}^3$ **3.** 30 in.^3, 24 in.^3, 150 in.^3, 18 in.^3 **4.** $4 \text{ in.} \times 3 \text{ in.} \times 2 \text{ in.}$

5. Yes; it can fit 4 along the length, and 3 along the width and it is exactly high enough.

6. $3 \text{ in.} \times 2 \text{ in.} \times 5 \text{ in.}$; two layers of 6

Challenge: Skills and Applications

1. 6.4 cm **2.** 40 ft by 13 ft **3. a.** 70.69 in.^3

b. 29.23 in.^3 **c.** 116.91 in.^3 **4. a.** 31.42 ft^3

b. 12.28 ft^3 **c.** 53.89 ft^3

Lesson 12.5

Warm-Up Exercises

1. 729 units^3 **2.** $72\pi \approx 226.19 \text{ units}^3$

3. $72\pi \approx 226.19 \text{ units}^3$ **4.** 264 units^3

5. $16\sqrt{3} \approx 27.71 \text{ units}^3$

Daily Homework Quiz

1. 660 m^3 **2.** 1244.1 yd^3 **3.** 120 in.^3

4. $735\pi \text{ cm}^3$ **5.** 11 m

Lesson Opener

Allow 5 minutes.

1. *Sample answers:* **a.** pyramid, prism **b.** cone, cylinder

c. Pyramid, prism, cone, cylinder

2.
Volume of one pyramid $= \frac{1}{3} \cdot$ Volume of the cube

3. *Sample answers:*

Volume of pyramid $= \frac{1}{3} \cdot$ Volume of prism;

Volume of cone $= \frac{1}{3} \cdot$ Volume of cylinder

Practice A

1. $9\sqrt{3} \approx 15.6 \text{ units}^2$ **2.** 64 units^2

3. $16\pi \approx 50.3 \text{ units}^2$ **4.** 120 in.^3

5. $170\frac{2}{3} \text{ cm}^3$ **6.** 336 cm^3 **7.** $50\sqrt{3} \approx 86.6 \text{ ft}^3$

8. $\dfrac{825\sqrt{3}}{4} \approx 357.24 \text{ in.}^3$

9. $144\sqrt{3} \approx 249.42 \text{ cm}^3$

10. 100.53 in.^3 **11.** 287.98 cm^3

12. 75.40 mm^3 **13.** 79.52 cm^3 **14.** $\approx 6.77 \text{ ft}$

Practice B

1. $121.5\sqrt{3} \approx 210.44 \text{ in.}^2$ **2.** $4\sqrt{3} \approx 6.93 \text{ cm}^2$

3. $36\pi \approx 113.10 \text{ in.}^2$ **4.** 384 in.^3

5. $\approx 32.48 \text{ in.}^3$ **6.** $\approx 127.31 \text{ cm}^3$

7. 201.06 cm^3 **8.** 163.49 cm^3 **9.** 1609.40 in.^3

10. The volume of concrete is about 150.8 cubic yards which is adequate to finish the job.

11. $6{,}250{,}000 \text{ tons}$ **12.** 682.7 cm^3

13. 679.4 in.^3

Practice C

1.   **2.**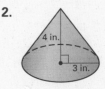

$V = 25 \text{ cm}^3$ $V \approx 37.7 \text{ in.}^3$

3. 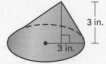 **4.**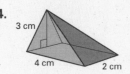

$V \approx 28.3 \text{ in.}^3$ $V \approx 8 \text{ cm}^3$

5. 27 cm **6.** 4.5 cm **7.** 11.5 in.

Lesson 12.5 *continued*

8. No, each cone would require about 5.06 grams of gold. For all twelve the jeweler would need about 60.7 grams. **9.** 1840 in.^{3} **10.** $533\frac{1}{3}$ m^3
11. 963.4 cm^3 **12.** 24 in.^{3}

Reteaching with Practice

1. 245 cm^3 **2.** 80 in.^{3} **3.** 10.4 ft^3
4. 251.3 in.^{3} **5.** 1400.4 m^3 **6.** 3451 m^3
7. 4.8 in. **8.** 4 m **9.** 6 m

Interdisciplinary Application

1. 85,730,400 ft^3 **2.** 36 ft^3 **3.** 82,800,000 ft^3
4. 2,930,400 ft^3

Challenge: Skills and Applications

1. 36 in.^{3} **2.** 180 cm^3

3. a. *Sample answer:*

$$V = \left(\begin{array}{c}\text{volume of} \\ \text{large pyramid}\end{array}\right) - \left(\begin{array}{c}\text{volume of} \\ \text{small pyramid}\end{array}\right)$$

$$= \frac{1}{3}B(h + k) - \frac{1}{3}Ck$$

$$= \frac{1}{3}Bh + \frac{1}{3}(B - C)k$$

b. *Sample answer:* $\left(\dfrac{h + k}{k}\right)^2 = \dfrac{B}{C}$,

$$\frac{h + k}{k} = \sqrt{\frac{B}{C}}, \quad \frac{h}{k} + 1 = \sqrt{\frac{B}{C}}$$

c. *Sample answer:* $\dfrac{h}{k} + 1 = \sqrt{\dfrac{B}{C}}$,

$$\frac{h}{k} = \frac{\sqrt{B} - \sqrt{C}}{\sqrt{C}},$$

$$\frac{k}{h} = \frac{\sqrt{C}}{\sqrt{B} - \sqrt{C}} \cdot \frac{\sqrt{B} + \sqrt{C}}{\sqrt{B} + \sqrt{C}}$$

$$= \frac{\sqrt{BC} + C}{B - C} = \frac{C + \sqrt{BC}}{B - C}$$

d. *Sample answer:*

$$V = \frac{1}{3}Bh + \frac{1}{3}(B - C)k$$

$$= \frac{1}{3}Bh + \frac{1}{3}(B - C) \cdot \frac{C + \sqrt{BC}}{B - C}h$$

$$= \frac{1}{3}Bh + \frac{1}{3}\left(C + \sqrt{BC}\right)h$$

$$= \frac{h}{3}\left(B + C + \sqrt{BC}\right)$$

4. a. 78 ft^3 **b.** $k = 2$; 78 ft^3

Quiz 2

1. 468.75 in.^{3} **2.** 2000 cm^3 **3.** 21,205.8 m^3
4. $12\pi \approx 37.7$ in.^{3} **5.** 982.4 ft^3 **6.** 512 cm^3
7. 80 cm^3

Lesson 12.6

Warm-Up Exercises

1. $400\pi \approx 1257$ units^3

2. $216\pi \approx 678.6$ units^3

3. $360\pi \approx 1131$ units^2 **4.** $55\pi \approx 172.8$ units^2

Daily Homework Quiz

1. $80\sqrt{3} \approx 138.56$ in.^{3} **2.** 252.89 ft^3
3. 18 m **4.** 192 cm^3

Lesson Opener

Allow 15 minutes.

1. *Sample answer:* Calculations on a spreadsheet may vary slightly.

Planet	Surface area (mi^2)	S.A. of planet / S.A. of Earth
Mercury	29,000,000	0.15
Venus	177,000,000	0.90
Earth	197,000,000	1
Mars	56,000,000	0.28
Jupiter	24,689,000,000	125.64
Saturn	17,545,000,000	89.28
Uranus	3,156,000,000	16.06
Neptune	2,963,000,000	15.08
Pluto	6,000,000	0.03

2. *Sample answer:* Mercury has about 15% of the surface area of Earth. Venus has about 90% of the surface area of Earth. Mars has about 28% of the surface area of Earth. Jupiter has about 126 times the surface area of Earth. Saturn has about 89 times the surface area of Earth. Uranus has about 16 times the surface area of Earth. Neptune has about 15 times the surface area of Earth. Pluto has about 3% of the surface area of Earth.

Lesson 12.6 *continued*

Technology Activity

1. The surface area is quadrupled and the volume is multiplied by eight.

2. The surface area is multiplied by 25 and the volume is multiplied by 125.

3. n^2 ; n^3

Practice A

1. $\overline{ED}$ **2.** $\overline{CA}$ **3.** $\overline{AB}$ **4.** ≈ 18.85 in.

5. $36\pi \approx 113.10$ in.² **6.** $36\pi \approx 113.10$ in.³

7. 201.06 cm² **8.** 380.13 in.² **9.** 452.39 cm²

10. hemisphere **11.** 4.5 in. **12.** 9 in.

13. ≈ 127.23 in.² **14.** 1436.76 in.³

15. 555.65 cm³ **16.** 8.18 in.³ **17.** The surface area of Mercury is about $\frac{1}{7}$ that of Earth's surface area.

Practice B

1. $S = 9\pi$ in.² ≈ 28.3 in.², $V = \frac{9}{2}\pi$ in.³ ≈ 14.1 in.³

2. $S = 100\pi \approx 314.2$ cm², $V = \frac{500}{3}\pi \approx 523.6$ cm³

3. $S = 7.84\pi$ m² ≈ 24.6 m², $V = 3.658\overline{6}\pi$ m³ ≈ 11.5 m³

4. 20π mm, 400π mm², $\frac{4000}{3}\pi$ mm³

5. 18 in., 1296π in.², 7776π in.³

6. 24 cm, 48π cm², $18{,}432\pi$ cm³

7. 5 yd, 10π yd, 100π yd²

8. $S = 452.39$ cm², $V = 469.14$ cm³

9. $S = 89.21$ in.², $V = 58.64$ in.³

10. $S \approx 128.68$ cm², $V \approx 137.26$ cm²

11. ≈ 205.89 cm³ **12.** $\approx 138{,}151{,}641$ mi²

13. $\approx 2.60711883 \times 10^{11}$ mi³

Practice C

1. $S = 153.94$ cm², $V = 179.59$ cm³

2. $S = 706.86$ in.², $V = 1767.15$ in.³

3. $S = 168.95$ ft², $V = 206.49$ ft³

4. 30π cm, 900π cm², 4500π cm³

5. 13 ft, 676π ft², $\frac{8788}{3}\pi$ ft³

6. 32 in., 64π in., $\frac{131{,}072}{3}\pi$ in.³

7. 7 yd, 14π yd, 196π yd² **8.** π cm²

9. 16π in.² **10.** 16π m²

11. $S \approx 27{,}646$ ft², $V \approx 385{,}368.7$ ft³

12. $\approx 41{,}787.16$ cm³ **13.** ≈ 52 cm³

Reteaching with Practice

1. 804.2 cm² **2.** 804.2 m² **3.** 1633.1 ft²

4. 28.3 m² **5.** 31.8 ft² **6.** 3.1 in.²

7. 44.6 m³ **8.** 1436.8 in.³ **9.** 904.8 cm³

Real-Life Application

1. radius ≈ 31.72 ft; diameter ≈ 63.44 ft, circumference ≈ 199.30 ft **2.** $\approx 12{,}643.76$ ft²

3. radius is approximately 25.18 feet; about 20% reduction in radius compared to a 50% reduction in volume

Challenge: Skills and Applications

1. $4r$ **2. a.** 904.78 cm³ **b.** $V_{\text{bead}} = \frac{1}{6}\pi h^3$

3. a. The bead is formed by removing the slice of the sphere that contains the liquid, a second slice congruent to the first, and the cylinder with radius r. The diameter of the entire sphere is $2R$ and the height of each slice is d, so the height of the bead is $2R - d - d$, or $2R - 2d$.
$V = \frac{4}{3}\pi(R - d)^3$ or
$\frac{4}{3}\pi R^3 - 4\pi R^2 d + 4\pi R d^2 - \frac{4}{3}\pi d^3$

b. leg: $R - d$; $r = \sqrt{2Rd - d^2}$;
$V = 2\pi(2Rd - d^2)(R - d)$, or
$4\pi R^2 d - 6\pi R d^2 + 2\pi d^3$

c. $V = \dfrac{1}{2}\left[\dfrac{4}{3}\pi R^3 - (4\pi R^2 d - 6\pi R d^2 + 2\pi d^3) - \left(\dfrac{4}{3}\pi R^3 - 4\pi R^2 d + 4\pi R d^2 - \dfrac{4}{3}\pi d^3\right)\right]$
$= \dfrac{1}{2}\left(2\pi R d^2 - \dfrac{2}{3}\pi d^3\right) = \dfrac{\pi d^2}{3}(3R - d)$

4. a. $V = 0$;

The volume makes sense because if the depth $= 0$, there is no liquid.

b. $V = \frac{2}{3}\pi R^3$;

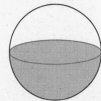

The volume makes sense because the depth is the radius of the sphere and the volume is half the volume of the sphere.

Lesson 12.6 *continued*

c. $V = \frac{4}{3}\pi R^3$;

The volume makes sense because the depth is the diameter of the sphere and the volume is equal to the volume of the sphere.

5. $\frac{5}{32}$; the result is reasonable because although the depth is one-fourth the diameter of the sphere, the volume of the liquid should be less than one-fourth the volume of the sphere because it is at the bottom of the sphere, which is not as wide as the middle.

Lesson 12.7

Warm-Up Exercises

1. 10 units^2, 2 units^3 **2.** 90 units^2, 54 units^3
3. 24π units^2, 12π units^3

Daily Homework Quiz

1. 314.16 m^2 **2.** 448.92 ft^3
3. 36π cm^2, 36π cm^3
4. 4; $16\pi \approx 50.27$ sq. units

Lesson Opener

Allow 10 minutes.

1. The volume is the same as the number of sugar cubes used.

Side length (units)	Surface area (sq units)	Volume: (cubic units)
1	6	1
2	24	8
3	54	27
4	96	64
5	150	125

2. Surface area $= 6 \cdot n^2$ **3.** Volume $= n^3$

4. The surface area increases by a factor of n^2. The volume increases by a factor of n^3.

Practice A

1. not similar   **2.** yes; 3:2   **3.** 3:1   **4.** 3:1
5. 3:1   **6.** 9:1   **7.** 27:1
8. $S = 832$ in.^{2}, $V = 1536$ in.^{3}

9. $S = 972\pi$ in.^{2}, $V = 2916\pi$ in.^{3}
10. $S = 2464$ cm^2, $V = 4096$ cm^3
11. $S = 202.5\pi$ cm^2, $V = 337.5\pi$ cm^3
12. 13,536 in.^{2}

Practice B

1. not similar   **2.** yes; 8:5
3. $S = 832\pi$ in.^{2}, $V = 2560\pi$ in.^{3}
4. $S = 3582$ in.^{2}, $V = 13,365$ in.^{3}
5. $S = 324\pi$ cm^2, $V = 972\pi$ cm^3
6. $S = 170.\overline{6}\pi$ cm^2, $V = 227.\overline{5}\pi$ cm^3 **7.** 1:48
8. ≈ 314.16 ft^2 **9.** ≈ 19.63 in.^{2}
10. ≈ 145.30 in.^{2}
11. $3200\pi \approx 10,053.10$ ft^3 **12.** ≈ 157.08 in.^{3}

Practice C

1. yes; 1:4   **2.** not similar
3. 256 in.^{2}, 224 in.^{3}
4. 2432π in.^{2}, $15,360\pi$ in.^{3} **5.** 324 cm^3, 1:3
6. 243 ft^2, 182.25 ft^3
7. $S = 90$ cm^2, $V = 50$ cm^3; $S = 810$ cm^2, $V = 1350$ cm^3 **8.** $S \approx 141.4$ in.^{2}, $V \approx 109.5$ in.^{3}; $S \approx 2262.4$ in.^{2}, $V \approx 7008$ in.^{3}
9. $S \approx 201.1$ ft^2, $V \approx 268.1$ ft^3; $S \approx 2463.4$ ft^2, $V \approx 11,494.8$ ft^3
10. 5.8 ft high, 5.5 ft wide; and 12.8 ft long
11. 1.56 ft^3

Reteaching with Practice

1. yes; 2:3   **2.** no   **3.** yes; 1:3
4. 339.3 cm^2; 477.1 cm^3 **5.** 328 ft^2; 336 ft^3

Cooperative Learning Activity

1. Answers may vary.   **2.** Answers may vary.
3. 9:1   **4.** 27:1

Interdisciplinary Application

1. 1:24   **2.** 30 ft $\times$ 40 ft   **3.** 156 ft^2
4. 160 ft^3

Challenge: Skills and Applications

1. 4.58 in.; 101.90 in.^{2} **2.** 27   **3.** 64
4. 7   **5.** 3, 5   **6.** 187.5 g
7. a. 4:5   **b.** 500 cm^2

Review and Assessment

Review and Assessment

Review Games and Activities

1. edge 2. octahedron 3. radius 4. cone
5. prism 6. dodecahedron 7. similar solids
8. volume 9. lateral surface
10. Platonic solids 11. sphere
12. Cavalieri's Principle 13. cylinder
Famous Mathematician: Leonhard Euler

Test A

1. 5, 5, 8 2. 9, 14, 21 3. 8, 12, 18
4. square 5. triangle 6. 294 in.^{2}
7. 251.33 cm^2 8. 12.6 m^2, 66.6 m^2
9. 27.4 ft^2, 158.6 ft^2 10. 314.2 ft^3, 282.7 ft^2
11. 68.6 m^3, 104.1 m^2 12. 216 in.^{3}
13. 360 ft^3 14. 15 cm^3 15. 6082.12 m^3
16. 203.58 in.^{3} 17. 12 in.^{3} 18. 28.87 ft^3
19. 164.64 m^3, 145.27 m^2 20. 20,579.53 in.^{3},
3631.68 in.^{2} 21. not similar 22. similar

Test B

1. 8, 6, 12 2. 7, 10, 15 3. 8, 12, 18
4. hexagon 5. rectangle 6. 77.12 in.^{2}
7. 321.95 in.^{2} 8. 15.3 ft^2, 86.3 ft^2
9. 11.3 in.^{2}, 61.3 in.^{2} 10. 66.0 in.^{3}, 100.1 in.^{2}
11. 5.7 m^3, 19.5 m^2 12. 36 ft^3 13. 96 ft^3
14. 13.5 in.^{3} 15. 211.66 ft^3 16. 804.25 m^3
17. 32 cm^3 18. 218.24 ft^3 19. 3053.63 in.^{3},
1017.88 in.^{2} 20. 321.56 m^3, 226.98 m^2
21. similar 22. not similar

Test C

1. 9, 9, 16 2. 8, 12, 18 3. 12, 20, 30
4. triangle 5. hexagon 6. 139.2 cm^2
7. 296.88 cm^2 8. 20.1 in.^{2}, 116.5 in.^{2}
9. 5.8 cm^2, 29.6 cm^2 10. 5.7 m^3, 19.5 m^2
11. 37.7 ft^3, 75.4 ft^2 12. 201.6 cm^3 13. 25 ft^3
14. 232 ft^3 15. 211.66 m^3 16. 753.98 ft^3
17. 384 in.^{3} 18. 609.7 cm^3
19. 321,555.10 cm^3, 22,698.01 cm^2
20. 523.60 ft^3, 314.16 ft^2

21. not similar 22. similar

SAT/ACT Test

1. C 2. B 3. E 4. A 5. C 6. D 7. B
8. E 9. A 10. C

Alternative Assessment

1. Complete answers should include:
• The number of faces (F), vertices (V), and edges (E) of a polyhedron are related by the formula $F + V = E + 2$. **a.** Edges = 12
b. Vertices = 8 **c.** Faces = 19
2. **a.** $V = 9\pi \approx 28.27$ cubic inches
b. Answers may vary. Answers should suggest the diameter decreasing. **c.** $d \approx 2.68$ inches
d. $h \approx 6$ inches **3.** The old can uses about 37.70 square inches. The new can uses about 47.12 square inches. No, they do not use the same amount of paper. The new can will use more.

4. Answers may vary. Answers should show that none of the cans are similar. Diagrams should be included.

Project: Spherical Modeling

Spherical shape	cuboctahedron	tetrahedron	cube	octahedron
Number of edges	24	6	12	12
Number of vertices	12	4	8	6
Number of faces	14	4	6	8

Students should discover that Euler's Formula holds for spherical polyhedra.

Comparisons of different models will vary. The following are some possible comparisons based on several types of characteristics. • Types of faces: Tetrahedron and octahedron use only spherical triangles. Cube uses only spherical squares. Cuboctahedron uses both. • Number of shapes meeting at a vertex: Tetrahedron and cube have 3 shapes meeting at each vertex. Octahedron and cuboctahedron have 4 shapes meeting at each vertex. • Size and number of faces: The models with more faces use smaller spherical triangles or squares for each face. Cuboctahedron (14 faces) has the smallest faces. Octahedron (8 faces) and cube (6 faces) have medium-sized faces. Tetrahedron (4 faces) has the largest faces.

Review and Assessment *continued*

• Vertices and faces in cube and octahedron: These numbers are reversals of each other. The cube has 8 vertices and 6 faces and the octahedron has 6 vertices and 8 faces.   • Great circles: Only the octahedron and the cuboctahedron contain great circles that go all the way around the widest part of the sphere. The octahedron has 3 great circles. If 1 great circle is thought of as the equator, the other 2 great circles go through the north and south poles. The cuboctahedron has 4 great circles. If 1 is thought of as the equator, the other 3 form the three sides of triangles that surround the north and south poles, but the great circles do not actually go through the poles. The other models do not contain any great circles.

• Symmetry: For some polyhedra, the top and bottom halves are symmetric if the model rests on a face. This is true for the cube and for the cuboctahedron if is rests on a square face (but not if it rests on a triangular face). For some polyhedra, the top and bottom halves are symmetric if the model rests on a vertex. This is true for the octahedron and for the cuboctahedron. The top and bottom halves of the tetrahedron are not symmetric no matter how the model is placed.

Cumulative Review

1. HL Congruence Theorem **2.** ASA Congruence Postulate or AAS Congruence Theorem **3.** $\overline{BC}, \overline{AB}, \overline{AC}$ **4.** $\overline{JL}, \overline{LK}, \overline{JK}$

5. 20 **6.** 3 **7.** $x = 5, y = 13$

8. $x = 120, y = 6$ **9.** $x = 14.3, y = 11.9$

10. $x = 10.9, y = 5.1$ **11.** 167.55 cm^2

12. 15.45 ft^2 **13.** 8.70 in.^2 **14.** 6 **15.** 12

16. 402.12 cm^2 **17.** 282.74 ft^2 **18.** 250 in.^3

19. 1568 in.^3 **20.** $268.08 \text{ in.}^3, 201.06 \text{ in.}^2$

21. $1.77 \text{ ft}^3, 7.07 \text{ ft}^2$